DETERMINISTIC OBSERVATION THEORY AND APPLICATIONS

This book presents a general theory as well as a constructive methodology to solve "observation problems," that is, reconstructing the full information about a dynamical process on the basis of partial observed data. A general methodology to control processes on the basis of the observations is also developed. Illustrative but also practical applications in the chemical and petroleum industries are shown.

This book is intended for use by scientists in the areas of automatic control, mathematics, chemical engineering, and physics.

J-P. Gauthier is Professor of Mathematics at the Université de Bourgogne, Dijon, France.

I. Kupka is Professor of Mathematics at the Université de Paris VI, France.

DETERMINISTIC OBSERVATION THEORY AND APPLICATIONS

JEAN-PAUL GAUTHIER

IVAN KUPKA

CAMBRIDGE
UNIVERSITY PRESS

CAMBRIDGE UNIVERSITY PRESS
Cambridge, New York, Melbourne, Madrid, Cape Town, Singapore,
São Paulo, Delhi, Dubai, Tokyo, Mexico City

Cambridge University Press
The Edinburgh Building, Cambridge CB2 8RU, UK

Published in the United States of America by Cambridge University Press, New York

www.cambridge.org
Information on this title: www.cambridge.org/9780521183864

First published 2001
First paperback edition 2010

A catalogue record for this publication is available from the British Library

Library of Congress Cataloguing in Publication data

Gauthier, Jean-Paul.

Deterministic observation theory and applications / Jean-Paul Gauthier, Ivan Kupka.

p. cm.

Includes bibliographical references and index.

ISBN 0-521-80593-7

1. Observers (Control theory) 2. Missing observations (Statistics) I. Kupka, Ivan, 1937– II. Title.

QA402.3 .G378 2001

003–dc21 2001025571

ISBN 978-0-521-80593-3 Hardback
ISBN 978-0-521-18386-4 Paperback

We dedicate this book to our wives, Irène and Prudence, respectively

The purpose of this book is to present a complete theory of observability and observation of finite dimensional nonlinear systems in the deterministic setting. The theory is used to prove very general results in dynamic output stabilization of nonlinear systems. Two real concrete applications are briefly described.

Dijon, September 9, 2000

Contents

Preface

A long time ago, while working on paper [19], we felt that there was a need to write a book on the subject of observability. Now, after many vicissitudes, this is a done thing.

During the conception of the book, the very novel point of view we had developed in our papers did not change. We discovered that it was really the right one, and was extremely efficient. In fact, based on it, we could build a totally new, complete, and general theory, and a new methodology for the problems related to observability, such as "output stabilization." At the same time, we applied our methodology to practical problems, and we realized that our methods were extremely efficient in practice.

At the very beginning, we intended to write a "survey" on the problems of observability, including nonlinear filtering. As the work progressed, we changed our minds. First, from the practical point of view, we faced a daunting task: a book of that type, had it ever seen the light of a day, would have been a monster. But, more important, our theory would have been drowned in a mass of disparate, disconnected facts.

Hence, this book presents only the general theory we have discovered, with a selection of real-life applications to convince the reader of the practical capability of the method. We strictly avoided the type of academic examples which are rife in many control theory publications.

Several principles guided us in the elaboration of this book:

- First, the book should be short. Including some developments in the stochastic context was a definite possibility, but this would have required the use of deep mathematical tools for meager returns. Enough mathematical theories already are used in the book.
- Second, the book is an excellent opportunity to convince people with a mathematical bent that "observation theory" is not out of place in mathematics. For that reason, the style of this book is a mathematical one. Also, we want to show that applied problems in the real world

can be dealt with by using beautiful mathematics. On the other hand, mathematics is not the main object of this book, but an excellent tool to achieve our goals.

- Third, we want to convince applied people (e.g., control engineers, chemical engineers) that our methodology is efficient. Therefore, they should strive to understand it and, above all, to use it. For this purpose, we want to point out the following:

    - We strove to make all the necessary mathematical tools accessible to uninitiated readers.
    - Bypassing the details of the proofs does not impair the understanding of the statements of the theoretical results and the constructive parts of the theory (many of the proofs are not obvious).
    - Chapter 8, containing the applications, is friendlier to the nonmathematical reader, albeit rigorous.

The development of the practical applications in this book, and of others not mentioned in it, was possible thanks to the cooperation with the French branch of the Shell company, its research center at "Grand-Couronne." One of the applications was actually implemented at the refinery of Petit-Couronne (France). The first author particularly wants to express his deep gratitude to the whole process control group there, more especially to Denis Bossane, François Deza, Marjoleine Van Doothing, and Frederic Viel, for their help, support, and for the good time we spent together. A very special and friendly remembrance goes to Daniel Rakotopara, the head of the group at that time, who so unfortunately died recently.

J-P. Gauthier expresses his warmest thanks to Jean-Jacques Dell'amico (head of the research center) and Pierre Sommelet (chief of the group), who not only took care of financial needs, but also are great friends.

Chapters 3, 4, 5, and 7 of this book contain, among others, the results of the papers [18], [19], [32]. For their kind permission to reproduce parts of papers [18], [19], and [32], we thank, respectively, the Society for Industrial and Applied Mathematics (Observability and observers for nonlinear systems, *SIAM Journal on Control*, Vol. 32, No. 4, pp. 975–994, 1994), Springer-Verlag (Observability for systems with more outputs than inputs, *Mathematische Zeitschrift* 223, pp. 47–78, 1996), and Kluwer Academic Publishers (with P. Jouan, Finite singularities of nonlinear systems. Output stabilization, observability, and observers. *Journal of Dynamical and Control Systems* 2(2), pp. 255–288, 1996).

Mexico City, September, 2000

1

Introduction

In this book, we present a **new, general, and complete theory of observability** and observation, deriving from our papers [18, 19, 32]. This theory is entirely **in the deterministic setting**. Let us mention here that there are several papers preceding these three that exploit the same basic ideas with weaker results. See [16, 17], in collaboration with H. Hammouri.

A list of all main notations is given in an index, page 221.

1. Systems under Consideration

We are concerned with general nonlinear systems of the form:

$$(\Sigma)\quad \begin{aligned} \frac{dx}{dt} &= f(x,u), \\ y &= h(x,u), \end{aligned} \tag{1}$$

typically denoted by Σ, where x, **the state**, belongs to X, an n-dimensional, connected, Hausdorff paracompact differentiable manifold, y, the **output**, takes values in R^{d_y}, and u, the **control variable**, takes values in $U \subset R^{d_u}$. For the sake of simplicity, we take $U = R^{d_u}$ or $U = I^{d_u}$, where $I \subset R$ is a closed interval. But typically U could be any closed submanifold of R^{d_u} with a boundary, a nonempty interior, and possibly with corners. Unless explicitly stated, X has no boundary.

The set of systems will be denoted by $S = F \times H$, where F is the set of u-parametrized vector fields f, and H is the set of functions h. In general, **except when explicitly stated**, f and h are C^∞. However, depending on the context, we will have to consider also analytic systems (C^ω), or C^r systems, for some $r \in \mathcal{N}$. Thus, if necessary, the required degree of differentiability will be stated, but in most cases the notations will remain S, F, H.

The simplest case is when U is empty, the so-called "uncontrolled case." In that situation, we will be able to prove more results than in the general case.

Usually, in practical situations, the output function h of the system does not depend on u. Unfortunately, from the theoretical point of view, this assumption is very awkward and leads to clumsy statements. For that reason, we will currently assume that h depends on the control u.

2. What Is Observability?

The preliminary definition we give here is the oldest one; it comes from the basic theory of linear control systems.

Roughly speaking, "observability" stands for the possibility of reconstructing the full trajectory from the observed data, that is, from the output trajectory in the uncontrolled case, or from the couple (output trajectory, control trajectory) in the controlled case. In other words, observability means that the mapping

$$initial - state \to output - trajectory$$

is injective, for all fixed control functions. More precise definitions will be given later in the book.

3. Summary of the Book

1. When the number d_y of observations is smaller than or equal to the number d_u of controls, then the relevant observability property is very rigid and is not stable under small perturbations, for germs of systems. Because of that rigidity, this observability property can be given a simple geometric characterization. This is the content of the paper [18] and the purpose of Chapter 3.

2. If, on the contrary, $d_y > d_u$, a remarkable phenomenon happens: The observability becomes generic, in a very strong sense, and for very general classes of control functions.

In Chapter 4, we state and prove a cornucopia of genericity results about observability as we define it. The most important of these results are contained in paper [19]. Some of these results present real technical difficulties.

3. The singular case: in the preceding two cases, the *initial – state → output – trajectory* mapping is regular. What happens if it becomes singular? This problem is too complex. In classical singularity theory, there is a useful and manageable concept of mapping with singularities: that of a "finite mapping." It is interesting that, in the uncontrolled analytic case, this concept can be extended to our *initial – state → output – trajectory* mappings, according to a very original idea of P. Jouan. This idea leads to the very interesting results of paper [32]. The controlled case is very different: If the system is singular, then it is not controllable. In this case, we also have several results,

giving a complete solution of the observation problem. These developments form the content of Chapter 5.

4. Observers: An observer is a device that performs the practical task of state reconstruction. In all cases mentioned above, $(1, 2, 3)$, an asymptotic observer can be constructed explicitly, under the guise of a differential equation that estimates the state of the system asymptotically. The estimation error has an arbitrarily large exponential decay. This is the so-called "high gain construction."

This construction is an adaptation to the nonlinear case of the "Luenberger," or of the "extended Kalman filter" method. The last one performs very well in practice. We will present these topics in Chapter 6.

5. Output stabilization: this study can be applied to output stabilization in the preceding cases 1, 2, and 3 above. One of our main results in [32] states that one can stabilize asymptotically a system via an asymptotic observer, using the output observations only, if one can stabilize it asymptotically using smooth state feedback.

This result is "semi-global": One can do this on arbitrarily large compacta. Let us note that, in cases 1 and 2, the $initial - state \rightarrow output - trajectory$ mapping is always immersive. In that case, the stabilizing feedback can be arbitrary. But, if the $initial - state \rightarrow output - trajectory$ mapping is not immersive, then it has to belong to a certain special ring of functions. These results are developed in Chapter 7.

6. In the last chapter, Chapter 8, we give a summary description of two applications in the area of chemical engineering. These represent the fallout from our long cooperation with the Shell company.

The first one, about distillation columns, is of practical interest because distillation columns really are **generic** objects in the petroleum and chemical industries. This application is a perfect illustration of the methods we are proposing for the problems of both observation and dynamic output stabilization.

The second application deals with polymerization reactors, and it constitutes also a very interesting and pertinent illustration. Both applications are the subjects of the articles [64, 65].

The classical notions of observability are inadequate for our purposes. For reasons discussed in the next section, Chapter 2 is devoted to the introduction of new concepts of observability. We hope that our book will vindicate our iconoclastic gesture of discarding the old observability concepts.

4. The New Observability Theory Versus the Old Ones

As we have said, observability is the injectivity of the mapping: $initial - state \rightarrow output - trajectory$. However, the concept of injectivity per se is very hard to handle mathematically because it is unstable. Hence, we have to introduce stronger concepts of observability, for example adding to the injectivity the condition of immersivity (infinitesimal injectivity), as in the classical theory of differentiable mappings.

In this book, we haven't discussed any of the other approaches to observability that have been proposed elsewhere, and we haven't referenced any of them. The reason for this is simple: We have no use for either the concepts nor the results of these other approaches.

In fact, we claim that our approach to observability theory, which is entirely new, is far superior to any of the approaches proposed so far.

Since we cannot discuss all of them, let us focus on the most popular: **the output injection method.**

The **output injection method** is in the spirit of the **feedback linearization** method (popular for the control of nonlinear systems). As for the feedback linearization, one tries to go back to the well-established theory of linear systems. First, one characterizes the systems that can be written as a linear system, plus a perturbation depending on the outputs only (in some coordinates). Second, for these systems only, one applies slight variations of the standard linear constructions of observer systems. This approach suffers from terminal defects.

A. It applies to an extremely small class of systems only. In precise mathematical terms, it means the following. In situation 2 above, where observability is generic, it applies to a class of systems of infinite codimension. In case 1, where the observability is nongeneric, it also applies to an infinite codimension subset of the set of observable systems.
B. Basically, the approach ignores the crucial distinction between the two cases: 1. $d_y \leq d_u$, 2. $d_y > d_u$.
C. The approach does not take into account generic singularities, and it is essentially local in scope.

Of course, these defects have important practical consequences in terms of sensitivity. In particular, in case 2, where the observability property is stable, the method is unstable.

5. A Word about Prerequisites

In this book, we have tried to keep the mathematical prerequisites to a strict minimum. What we need are the following mathematical tools: transversality theory, stratification theory and subanalytic sets, a few facts from several complex variables theory, center manifold theory, and Lyapunov's direct and inverse theorems.

For the benefit of the reader, a summary of the results needed is provided in the Appendix. It is accessible to those with only a modest mathematical background.

6. Comments

6.1. Comment about the Dynamic Output Stabilization Problem

At several places in the book, we make the assumption that the state space X, is just the Euclidean space R^n. If one wants only to estimate the state, this is not a reasonable assumption: the state space can be anything. However, for the dynamic output stabilization of systems that are state-feedback stabilizable, it is a reasonable assumption because the basin of attraction of an asymptotically stable equilibrium point of a vector field is diffeomorphic to R^n (see [51]).

6.2. Historical Comments

6.2.1. About "Observability"

The observability notion was introduced first in the context of linear systems theory. In this context, the **Luenberger observer**, and the **Kalman filter** were introduced, in the deterministic and stochastic settings, respectively.

For linear systems, **the observability notion is independent of the control function** (either the $initial - state \rightarrow output - trajectory$ mapping is injective for all control functions, or it is not injective for each control function). **This is no longer true for nonlinear systems.** Moreover, as we show in this book, in the general case where $d_y \leq d_u$, observability (for all inputs) is not at all a generic property. For these reasons (and certainly also just for tractability), several weaker different notions of observability have been introduced, which are generic and which agree with the old observability notion in the special case of linear systems. In this setting, there is the pioneer work [24]. As we said, these notions are totally inadequate for our purposes, and we just forget about them.

6.2.2. About Universal Inputs

Let us say that a control function **separates** two states, if the corresponding output trajectories, from these two initial states, do not coincide.

For a nonlinear system, a **universal input** is a control function that separates all the couples of states that can be separated by some control.

We want to mention a pioneer work by H. J. Sussmann [47], in which it is proved, roughly speaking, that "universal inputs do exist." For this purpose, the author made use of the properties of subanalytic sets, in a spirit very similar to the one in this book.

6.2.3. About the Applications

In Chapter 8, we present two applications from chemical engineering science. There are already several other applications of our theory in many fields, but we had to choose.

The two applications we have chosen look rather convincing, because they are not "academic," and some refinements of the theory are really used. Moreover, these two applications, besides their illustrative character, are very important in practice and have been addressed by research workers in control theory, using other techniques, for many years. It is hard to give an exhaustive list of other studies (related to control and observation theory) on distillation columns and polymerization reactors. However, let us give a few references that are significant:

For distillation columns: [58], [61], [62].
For polymerization reactors: [56], [57], [60], [59].

Regarding distillation columns, it would be very interesting (and probably very difficult) to study the case of **azeotropic distillations**, which is not addressed in this volume. It seems that all the theory collapses in this case of azeotropic distillations.

Part I

Observability and Observers

2

Observability Concepts

In this chapter, we will state and explain the various definitions of observability that will be used in this book (see Section 4 in Chapter 1).

1. Infinitesimal and Uniform Infinitesimal Observability

The space of control functions under consideration will just be the space $L^{\infty}[U]$ of all measurable bounded, U-valued functions $u : [0, T_u[\to U$, defined on semi-open intervals $[0, T_u[$ depending on u. The space of our output functions will be the space $L[R^{d_y}]$ of all measurable functions $y : [0, T_y[\to R^{d_y}$, defined on the semi-open intervals $[0, T_y[$. Usually, input and output functions are defined on closed intervals. However, this is irrelevant. The following considerations led us to work with semi-open intervals. For any input $\hat{u} \in L^{\infty}[U]$ and any initial state x_0, the maximal solution of the Cauchy problem for positive times

$$\frac{d\hat{x}}{dt} = f(\hat{x}(t), \hat{u}(t)), \hat{x}(0) = x_0$$

is defined on a semi-open interval $[0, e(\hat{u}, x_0)[$, where $0 < e(\hat{u}, x_0) \leq T_{\hat{u}}$. If $e(\hat{u}, x_0) < T_{\hat{u}}$, then, $e(\hat{u}, x_0)$ is the positive escape time of x_0 for the time dependent vector field $f(., \hat{u}(t))$. It is well known that, for all $\hat{u} \in L^{\infty}[U]$, the function $x_0 \to e(\hat{u}, x_0) \in \bar{R}_{+}^{*}$ is lower semi-continuous ($\bar{R}_{+}^{*} = \{a | 0 < a \leq \infty\}$).

Definition 1.1. The input-output mapping P of Σ is defined as follows:

$$P : L^{\infty}[U] \times X \to L[R^{d_y}], (\hat{u}, x_0) \to P(\hat{u}, x_0),$$

where $P(\hat{u}, x_0)$ is the function $\hat{y} : [0, e(\hat{u}, x_0)[\to R^{d_y}$ defined by

$$\hat{y}(t) = h(\hat{x}(t), \hat{u}(t)).$$

The mapping $P_{\hat{u}} : X \to L[R^{d_y}]$ is $P_{\hat{u}}(x_0) = P(\hat{u}, x_0)$.

Definition 1.2.[1] A system is called **observable** if for any triple $(\hat{u}, x_1, x_2) \in L^\infty[U] \times X \times X$, $x_1 \neq x_2$, the set of all $t \in [0, \min(e(\hat{u}, x_1), e(\hat{u}, x_2))[$ such that $P(\hat{u}, x_1)(t) \neq P(\hat{u}, x_2)(t)$ has positive measure.

Now, we define the "**first variation**" of Σ, or the "**lift of Σ on TX**." The mapping $f : X \times U \to TX$ induces the partial tangent mapping $T_X f : TX \times U \to TTX$ (tangent bundle of TX). Then, if ω denotes the canonical involution of TTX (see [1]), $\omega \circ T_X f$ defines a parametrized vector field on TX, also denoted by $T_X f$. Similarly, the function $h : X \times U \to R^{d_y}$ has a differential $d_X h : TX \times U \to R^{d_y}$. The first variation of Σ is the input–output system:

$$(T\Sigma)\begin{cases} \frac{d\xi}{dt} = T_X f(\xi, u) = T_X f_u(\xi), \\ \eta = d_X h(\xi, u) = d_X h_u(\xi). \end{cases} \tag{2}$$

Its input–output mapping is denoted by dP, and the trajectories of (1) and (2) are related as follows:

If $\xi : [0, T_\xi[\to TX$ is a trajectory of (2) associated with the input $\hat{u}$, the projection $\pi(\xi) : [0, T_\xi[\to X$ is a trajectory of Σ associated with the same input. Conversely, if $\varphi_t(x_0, \hat{u}) : [0, e(\hat{u}, x_0)[\to X$ is the trajectory of Σ starting from x_0 for the input $\hat{u}$, the map $x \to \varphi_\tau(x, \hat{u})$ is a diffeomorphism from a neighborhood of x_0 onto its image, for all $\tau \in [0, e(\hat{u}, x_0)[$. Let $T_X\varphi_\tau : T_{x_0}X \to T_z X$, $z = \varphi_\tau(x_0, \hat{u})$ be its tangent mapping. Then, for all $\xi_0 \in T_{x_0}X$:

$$e_{T\Sigma}(\hat{u}, \xi_0) = e_\Sigma(\hat{u}, \pi(\xi_0)) = e_\Sigma(\hat{u}, x_0),$$

and, for almost all $\tau \in [0, e(\hat{u}, x_0)[$:

$$dP(\hat{u}, \xi_0)(\tau) = d_X h(T_X\varphi_\tau(\hat{u}, \xi_0), \hat{u}(\tau)) = d_X\left(P^\tau_{\Sigma,\hat{u}}\right)(\xi_0). \tag{3}$$

The right-hand side of these equalities (3) is the differential of the function $P^\tau_{\Sigma,\hat{u}} : V \to R^{d_y}$, where V is the open set:

$$V = \{x \in X | 0 < \tau < e(\hat{u}, x)\}, \text{ and } P^\tau_{\Sigma,\hat{u}}(x) = P(\hat{u}, x)(\tau).$$

For any $a > 0$, let $L^\infty_{loc}([0, a[; R^{d_y})$ denote the space of measurable functions $v : [0, a[\to R^{d_y}$ which are locally in L^∞. For all $\hat{u} \in L^\infty(U)$, $x_0 \in X$, the restriction of dP to $\{\hat{u}\} \times T_{x_0}X$ defines a linear mapping:

$$\begin{aligned} dP_{\hat{u},x_0} &: T_{x_0}X \to L^\infty_{loc}([0, e(\hat{u}, x_0)[; R^{d_y}), \\ dP_{\hat{u},x_0}(\xi_0)(t) &= dP(\hat{u}, \xi_0)(t). \end{aligned} \tag{4}$$

[1] In nonlinear control theory, the notion of observability defined here is usually referred to as "uniform observability." Let us stress that it is just the old basic observability notion used for linear systems.

Definition 1.3. The system Σ is called infinitesimally observable at $(\hat{u}, x) \in L^\infty[U] \times X$ if the linear mapping $dP_{\hat{u},x}$ is injective. It is called infinitesimally observable at $\hat{u} \in L^\infty[U]$ if it is infinitesimally observable at all pairs $(\hat{u}, x)$, $x \in X$, and called **uniformly infinitesimally observable** if it is infinitesimally observable at all $\hat{u} \in L^\infty[U]$.

Remark 1.1. In view of the relation 3 above, the fact that the system is infinitesimally observable at $\hat{u} \in L^\infty[U]$ means that the mapping $P_{\hat{u}} : X \to L[R^{d_y}]$ is an immersion of X into $L[R^{d_y}]$ (as was stated, $P_{\hat{u}}$ is differentiable in the following sense: we know that $e(\hat{u}, x) \geq e(\hat{u}, x_0) - \varepsilon$ in a neighborhood U_ε of x_0. Then $P_{\hat{u}}$ is differentiable in the classical sense from U_ε into $L^\infty([0, e(\hat{u}, x_0) - \varepsilon]; R^{d_y})$. $P_{\hat{u}}$ is an immersion in the sense that these differential maps are injective).

This notion of uniform infinitesimal observability is the one which makes sense in practice, when $d_y \leq d_u$. In most of the examples from real life we know of, when $d_y \leq d_u$, the system is uniformly infinitesimally observable.

A very frequent situation in practice is the following: The physical state space for x is an open subset $\check{X}$ of X, and $\check{X}$ is positively invariant under the dynamics of Σ. The trajectories $\hat{x}(t)$ that are unobservable take their values in the boundary $\partial\check{X}$, and the corresponding controls $\hat{u}(t)$ take their values in ∂U. In particular, this will be the case for the first example we show in Chapter 8.

2. The Canonical Flag of Distributions

In this section, we assume that $d_y = 1$. As above, set: $h_u(x) = h(x, u)$, $f_u(x) = f(x, u)$. Associated with the system Σ, there is a family of flags $\{D_0(u) \supset D_1(u) \supset \ldots \supset D_{n-1}(u)\}$ of distributions on X (parametrized by **the value** $u \in U$ of the control). $D_0(u) = \text{Ker}(d_X h_u(x))$, where d_X denotes again the differential with respect to the x variable only. For $0 \leq k < n - 1$:

$$D_{k+1}(u) = D_k(u) \cap Ker\big(d_X\big(L_{f_u}^{k+1}(h_u)\big)\big),$$

where L_{f_u} is the Lie derivative operator on X, w.r.t. the vector field f_u. Let us set:

$$D(u) = \{D_0(u) \supset D_1(u) \supset \ldots \supset D_{n-1}(u)\}. \tag{5}$$

This u-dependent flag of distributions is not **regular** in general (i.e., $D_i(u)$ does not have the constant rank $n - i - 1$).

Definition 2.1. The flag $D(u)$ is called **the canonical flag** associated to Σ. In the case where the flag $D(u)$ is **regular and independent of** u (notation: $\partial_u D(u) = 0$), the canonical flag is said to be **uniform.**

The case in which $D(u)$ is uniform will be especially important in Chapter 3. In fact, this case will characterize uniform infinitesimal observability.

Note: Here, for us, a distribution D is just a subset of TX, the intersection of which with each tangent plane $T_x X$ is a vector subspace of $T_x X$. Once the flag $D(u)$ defined here is regular, the distributions $D_i(u)$ are smooth distributions in the usual sense.

3. The Phase-Variable Representation

Here $L^k_{f_u}(h_u)$ denotes the d_y-tuple of functions, the components of which are $L^k_{f_u}(h^i_u)$, where h^i_u is the ith component of $h_u = h(., u)$. We consider control functions that are sufficiently continuously differentiable only: k times, for example.

Consider $R^{(k-1)d_u} = R^{d_u} \times \ldots \times R^{d_u}$ ($k-1$ times) and $R^{kd_y} = R^{d_y} \times \ldots \times R^{d_y}$ (k times). We denote the components of $v \in R^{(k-1)d_u}$ by $(v', \ldots, v^{(k-1)})$ and the components of $y \in R^{kd_y}$ by $(y, y', \ldots, y^{(k-1)})$.

Definition 3.1. (Valid for X with corners.) There exist smooth mappings Φ^Σ_k and $S\Phi^\Sigma_k$ (the notation $S\Phi^\Sigma_k$ stands for **suspension** of Φ^Σ_k):

$$\begin{aligned} \Phi^\Sigma_k &: X \times U \times R^{(k-1)d_u} \to R^{kd_y}, \\ \Phi^\Sigma_k &: \left(x_0, u, u', \ldots, u^{(k-1)}\right) \to \left(y, y', \ldots, y^{(k-1)}\right), \end{aligned} \tag{6}$$

$$\begin{aligned} S\Phi^\Sigma_k &: X \times U \times R^{(k-1)d_u} \to R^{kd_y} \times R^{kd_u}, \\ S\Phi^\Sigma_k &: \left(x_0, u, u', \ldots, u^{(k-1)}\right) \to \left(y, y', \ldots, y^{(k-1)}, u, u', \ldots, u^{(k-1)}\right), \end{aligned} \tag{7}$$

which are polynomial in the variables $(u', \ldots, u^{(k-1)})$ and smooth in (x_0, u), such that if $(\hat{x}, \hat{u}) : [0, T_{\hat{u}}[\to X \times U$ is a semitrajectory of our system Σ starting at x_0, and $t \to y(t) = h(\hat{x}(t), \hat{u}(t))$ is the corresponding output trajectory. Then the jth derivative $y^{(j)}(0)$ of $y(t)$ at time 0 is the jth block-component of $\Phi^\Sigma_k(x_0, u(0), \frac{du}{dt}(0), \ldots, \frac{d^{k-1}u}{dt^{k-1}}(0))$.

Let us say that the system Σ has the **phase-variable property of order** k, denoted by $PH(k)$, if, for all $x_0 \in X$ and $u(.)$ k-times differentiable:

$$y^{(k)} = \breve{H}\left(S\Phi^\Sigma_k\left(x_0, u, u', \ldots, u^{(k-1)}\right), u^{(k)}\right), \tag{8}$$

for some smooth (C^∞) function $\breve{H} : R^{kd_y} \times R^{(k+1)d_u} \to R^{d_y}$. Notice that if such a function does exist, **it is not unique** in general.

If one denotes temporarily by C_k^∞ the ring of smooth functions $g : R^{kdy+(k+1)du} \to R$, the property $PH(k)$ means that the components $y_i^{(k)}$ of $y^{(k)}$ belong to the ring $\Re_k^\sharp$, pull back of C_k^∞ by the mapping $S\bar{\Phi}_k^\Sigma : \Re_k^\sharp = (S\bar{\Phi}_k^\Sigma)^* C_k^\infty$, where

$$S\bar{\Phi}_k^\Sigma = S\Phi_k^\Sigma \times Id^{d_u},$$
$$S\bar{\Phi}_k^\Sigma\left(x, u, u', \ldots, u^{(k-1)}, u^{(k)}\right) = \left(S\Phi_k^\Sigma\left(x, u, u', \ldots, u^{(k-1)}\right), u^{(k)}\right).$$

Then, we can consider the differential system Σ_k, on R^{kd_y}:

$$(\Sigma_k) \begin{cases} \dot{z}_1 = z_2, \ldots\ldots, \dot{z}_{k-1} = z_k, \\ \dot{z}_k = \breve{H}\left(z_1, \ldots, z_k, u(t), \ldots, \frac{d^k u}{dt^k}(t)\right). \end{cases} \tag{9}$$

This differential system Σ_k is called a **phase-variable representation** of Σ. It has the following property, a consequence of the uniqueness of the solutions of smooth O.D.E.s. For any C^k function u, Φ_k^Σ maps the trajectories $x(t)$ of Σ associated with $u(t)$ into the corresponding trajectories of Σ_k : If $x(t)$ is a trajectory of Σ corresponding to $u(t)$, then the curve $t \to \Phi_k^\Sigma(x(t), u(t), u'(t), \ldots, u^{(k-1)}(t))$ is the trajectory of Σ_k corresponding to $u(t)$, starting from $\Phi_k^\Sigma(x(0), u(0), u'(0), \ldots, u^{(k-1)}(0))$. In particular, the output trajectory $t \to y(t)$ is mapped into $t \to z_1(t)$, where z_1 denotes the first d_y components of the state z of Σ_k.

A very important particular case where the property $PH(k)$ holds is the following. Assume that the map $S\Phi_k^\Sigma$ is an **injective immersion.** For any open relatively compact subset $\Omega \subset X$, let us consider the restriction $S\Phi_k^{\Sigma,\Omega} = (S\Phi_k^\Sigma)_{|\Omega \times U \times R^{(k-1)d_u}}$. If I_Ω denotes the image of $S\Phi_k^{\Sigma,\Omega}$, $I_\Omega \subset R^{kd_y} \times R^{kd_u}$, then, $y^{(k)}(x, u, u', \ldots, u^{(k)})$ defines a function $\breve{h}$ on $I_\Omega \times R^{d_u}$ and easy arguments using partitions of unity show that we can extend this function smoothly to all of $R^{kd_y} \times R^{(k+1)d_u}$. We temporarily leave this simple fact for the reader to show: In Chapter 5, we will prove a slightly stronger (but not more difficult) result.

Denoting this extension of $\breve{h}$ by $\breve{H}$, we get a phase-variable representation of order k for Σ restricted to Ω.

As we shall see, there are other interesting cases in which the map $S\Phi_k^\Sigma$ is only injective, but $PH(k)$, the phase-variable property of order k, holds for Σ, for some k. This situation will be studied in Chapter 5.

Strongly related to this phase-variable property, are the notions of **differential observability**, and **strong differential observability**.

4. Differential Observability and Strong Differential Observability

"Differential observability" just means injectivity of the map $S\Phi_k^\Sigma$. "Strong differential observability" will mean that **moreover** $S\Phi_k^\Sigma$ is also an immersion. Let us relate precisely these notions to the notion of a **dynamical extension** of Σ.

The control functions u are assumed to be sufficiently smooth. We can consider the Nth dynamical extension Σ^N of Σ, and the Nth dynamical extension f^N of f, defined as follows. f^N is just the vector field on $X \times U \times R^{(N-1)d_u} = X \times U \times (R^{d_u} \times \ldots \times R^{d_u})$, ($N-1$ factors R^{d_u}),

$$f^N\left(x, u^{(0)}, \ldots, u^{(N-1)}\right) = \sum_{i=1}^{n} f_i\left(x, u^{(0)}\right)\frac{\partial}{\partial x_i} + \sum_{i=0}^{N-2}\sum_{j=1}^{d_u} u_j^{(i+1)}\frac{\partial}{\partial u_j^{(i)}}. \tag{10}$$

Moreover, if we set $b^N = (b_i^N)$, with $b_i^N = \frac{\partial}{\partial u_i^{(N-1)}}$, and $u^{(N)} = (u_i^{(N)})$, the **"new control variable,"**

$$u^{(N)}b^N = \sum_{i=1}^{d_u} u_i^{(N)} b_i^N,$$

then, we can give the following definition.

Definition 4.1. The Nth dynamical extension $\Sigma^N = (F^N, \check{h})$ of Σ, is just the control system on $X \times U \times R^{(N-1)d_u}$ with control variable $u^{(N)} \in R^{d_u}$, parametrized vector field $F^N = f^N + u^{(N)}b^N$, and observation function $\check{h} = (h(x, u^{(0)}), u^{(0)})$.

Remark 4.1. If $U = I^{d_u}$, $I \neq R$, the state space of Σ^N has corners.

In fact, Σ^N is just the system we get by adding to the state variables the $N-1$ first derivatives of the inputs. The Nth derivative is the new control. The observations are the observations of Σ and the control variables u_i denoted here by $u_i^{(0)}$, $1 \leq i \leq d_u$, to stress that the function is the zeroth derivative of itself. Also, $u^{(0)}$, (resp. $u^{(j)}$) denotes the vector with components $u_i^{(0)}$ (resp. $u_i^{(j)}$), $1 \leq i \leq d_u$.

Let us set $\underline{u}_N = (u^{(0)}, \ldots\ldots, u^{(N-1)})$, and more generally, for a smooth function $y(t)$, with successive derivatives $y^{(i)}(0)$ at $t = 0$, $\underline{y}_N = (y(0), y'(0), \ldots\ldots, y^{(N-1)}(0))$.

The maps Φ_N^Σ, $S\Phi_N^\Sigma$ have already been defined in Section 3. It will also be important to make the system Σ vary in the set S of systems. Hence, we will have to consider the following maps:

$$
\begin{aligned}
S\Phi_N &: X \times U \times R^{(N-1)d_u} \times S \to R^{Nd_y} \times R^{Nd_u},\\
(x, \underline{u}_N, \Sigma) &\to \left(h\left(x, u^{(0)}\right), L_{f^N} h(x, \underline{u}_2), \ldots, (L_{f^N})^{N-1} h(x, \underline{u}_N), \underline{u}_N\right)\\
&= (\underline{y}_N, \underline{u}_N) \qquad (11)\\
S\Phi_N^{\Sigma} &: X \times U \times R^{(N-1)d_u} \to R^{Nd_y} \times R^{Nd_u},\\
(x, \underline{u}_N) &\to S\Phi_N(x, \underline{u}_N, \Sigma),
\end{aligned}
$$

and

$$
\begin{aligned}
\Phi_N &: X \times U \times R^{(N-1)d_u} \times S \to R^{Nd_y},\\
(x, \underline{u}_N, \Sigma) &\to \left(h\left(x, u^{(0)}\right), L_{f^N} h(x, \underline{u}_2), \ldots, (L_{f^N})^{N-1} h(x, \underline{u}_N)\right) = \underline{y}_N,\\
\Phi_N^{\Sigma} &: X \times U \times R^{(N-1)d_u} \to R^{Nd_y}, \qquad (12)\\
(x, \underline{u}_N) &\to \Phi_N(x, \underline{u}_N, \Sigma).
\end{aligned}
$$

$$
S\Phi_N(x, \underline{u}_N, \Sigma) = (\Phi_N(x, \underline{u}_N, \Sigma), \underline{u}_N) = S\Phi_N^{\Sigma}(x, \underline{u}_N) = \left(\Phi_N^{\Sigma}(x, \underline{u}_N), \underline{u_N}\right).
$$

Definition 4.2. Σ is said to be **differentially observable** of order N, if $S\Phi_N^{\Sigma}$ is an injective mapping and to be **strongly differentially observable,** if $S\Phi_N^{\Sigma}$ is an injective immersion.

As we mentioned in Section 3, if Σ is strongly differentially observable of order N, then Σ possesses also the phase variable property $PH(N)$, when restricted to Ω, where Ω is any open relatively compact subset of X.

The reason for these definitions is that **when $d_y > d_u$, strong differential observability** is easily tractable for the purpose of construction of observer systems. Moreover, roughly speaking, **it is a generic property**. Therefore, it is a relevant definition in that case. This is the subject of Chapter 4. The motivation to consider differential (not strong) observability is that it is the most general concept adapted to the study of dynamic output stabilization (Chapter 7).

5. The Trivial Foliation

Associated to Σ, there is a subspace Θ^{Σ} of the space $C^{\infty}(X)$ of smooth functions $h : X \to R$.

The subspace Θ^{Σ} is the smallest subspace of $C^{\infty}(X)$ containing the components h_u^i of $h_u = h(., u)$, for all $u \in U$, which is closed under Lie differentiation on X, with respect to all of the vector fields $f_v = f(., v)$, $v \in U$. It is the real vector subspace of $C^{\infty}(X)$ generated by the functions $(L_{f_{u_r}})^{k_r}(L_{f_{u_{r-1}}})^{k_{r-1}} \ldots (L_{f_{u_1}})^{k_1}(h_{u_0})$, for $u_0, \ldots, u_r \in U$, where L denotes the Lie derivative operator on X.

Here, Θ^Σ is called the **observation space** of Σ. The space $d_X\Theta^\Sigma$ of differentials (w.r.t. x) of elements of Θ^Σ defines a codistribution that is, in general, singular. The distribution Δ_Σ annihilated by $d_X\Theta^\Sigma$ is called the **trivial distribution** associated to Σ. The level sets of Θ^Σ (i.e. the intersections of the level sets of elements of Θ^Σ) define the associated foliation, called the **trivial foliation**, associated to Σ.

These notions are classical [24]. We call this foliation the trivial foliation because it is actually trivial (in the sense that the leaves are zero-dimensional) for generic systems. This is also true for systems that are uniformly infinitesimally observable, as our results will show. However, **it is worth pointing out that the following theorem holds.**

Theorem 5.1. *The rank of* Δ_Σ *is constant along the (positive or negative time) trajectories of* Σ*, in the analytic case [24].*

Proof. By standard controllability arguments (recalled in Section 6 below), it is sufficient to show that the rank of Δ_Σ is the same at a point x_0 and at a point x_1 of the form,

$$x_1 = \exp t f_v(x_0),\ v \in U.$$

To show this, we use the following analytic expansions, for $\psi \in \Theta^\Sigma$, valid for small t:

$$\begin{aligned}(i)\ \psi(\exp(t\ f_v))(x_0) &= \sum_{k=1}^{\infty} \frac{t^k}{k!}(L_{f_v})^k \psi(x_0),\\ (ii)\ d_X(\psi(\exp(t\ f_v))(x_0)) &= \sum_{k=1}^{\infty} \frac{t^k}{k!} d_X((L_{f_v})^k \psi)(x_0).\end{aligned} \tag{13}$$

If we denote by T the tangent mapping of the mapping $\exp t\ f_v(x)$, $T : T_{x_0}X \to T_{x_1}X$, and by T^* its adjoint, Formula (13), (ii), shows that $T^* : T^*_{x_1}X \to T^*_{x_0}X$, maps the covector $d_X\psi(x_1) \in d_X\Theta^\Sigma(x_1)$ to a covector in $d_X\Theta^\Sigma(x_0)$. Hence, $rank(d_X\Theta^\Sigma(x_1)) \leq rank(d_X\Theta^\Sigma(x_0))$. Interchanging the role of x_0 and x_1, by taking $t = -t$, we conclude that $rank(\Delta_\Sigma(x_1)) = rank(\Delta_\Sigma(x_0))$. ■

Hence, as soon as the analytic system Σ is **controllable** in the weak sense of the transitivity of its Lie algebra (see Section 6), **the distribution** Δ_Σ **is regular**. In particular, the leaves of the trivial foliation (the level sets of Θ^Σ) are submanifolds of the same dimension.

Now, let us consider the case where Δ_Σ is regular, nontrivial, and not necessarily analytic. By Theorem 5.1, it is always regular in the analytic controllable case.

Exercise 5.1. Show that Δ_Σ is preserved by the dynamics of Σ. (i.e. for any control function $u(.) \in L^\infty[U]$, $T_X\varphi_t(., u)$ maps $\Delta_\Sigma(x_0)$ onto $\Delta_\Sigma(\varphi_t(x_0, u))$).

Because $h(., u)$ is constant on the leaves of Δ_Σ, for two distinct initial conditions, sufficiently close in the same leaf, the corresponding output trajectories coincide for t small enough, whatever the control function.

In particular, Σ is not observable, for **any fixed value of the control** function, even if restricted to small open sets: For each control, one can find couples of points, arbitrarily close, that are not distinguished by the observations, for small times.

The following simple fact is important. We leave it as an exercise.

Exercise 5.2. In the case $U = \varnothing$, show the following:

$$\begin{aligned}&\text{– If (iff in the } C^\omega \text{ case) } \Theta^\Sigma \text{ separates the points on } X,\\ &\text{then, } \Sigma \text{ is observable.}\end{aligned} \tag{14}$$

There is an alternative way to define the distribution Δ_Σ **in the analytic case**, which will be of interest, together with Theorem 5.1 in Chapter 5. Let us first define the vector subspace Ξ^Σ of $H = C^\infty(X \times U)$, as follows: Ξ^Σ is the smallest real vector subspace of H, which contains the components h^i of h, and that is closed under the action of the Lie derivatives L_{f_u} on X, and with respect to the derivations $\partial_j = \frac{\partial}{\partial u_j}$, $j = 1, \ldots, d_u$. Ξ^Σ is generated by functions of the form:

$$L_{f_u}^{k_1}(\partial_{j_1})^{s_1} L_{f_u}^{k_2}(\partial_{j_2})^{s_2} \ldots\ldots L_{f_u}^{k_r}(\partial_{j_r})^{s_r} h_{i,u}, \qquad k_i, s_i \geq 0. \tag{15}$$

Fixing $u \in U$, we obtain the vector subspace $\Xi^\Sigma(u) \subset C^\infty(X)$ and the space $d_X \Xi^\Sigma(u)$ of differentials of the elements of $\Xi^\Sigma(u)$, with respect to the x variable only. We call $\bar{\Delta}_\Sigma(u)$ the distribution annihilated by $d_X \Xi^\Sigma(u)$.

Theorem 5.2. *(a)* $\Delta_\Sigma \subset \bar{\Delta}_\Sigma(u)$; *(b) In the analytic case,* $\bar{\Delta}_\Sigma(u)$ *is independent of u and* $\Delta_\Sigma = \bar{\Delta}_\Sigma(u)$.

Proof. First, let us show that $\Delta_\Sigma \subset \bar{\Delta}_\Sigma(u)$, whatever u is.

Consider the following functions $\varphi_{i,k}(x, u, u_1, \ldots\ldots, u_r)$, $1 \leq i \leq d_y$:

$$\begin{gathered}\varphi_{i,k}(x, u, u_1, \ldots\ldots, u_r) = L_{f_{u+u_1}}^{k_1} \ldots\ldots L_{f_{u+\ldots+u_{r-1}}}^{k_{r-1}} h_{i,u+u_1+\ldots+u_r},\\ \varphi_{i,k} : D \to R,\ D \subset X \times U \times (R^{d_u})^r.\end{gathered} \tag{16}$$

As functions of x, for $u, u_1, \ldots, u_r$ fixed, these functions are generators of Θ^Σ. The main fact we will use is that

$$\frac{\partial}{\partial u_{1,i_1}}^{\alpha_{1,i_1}} \cdots \frac{\partial}{\partial u_{r,i_r}}^{\alpha_{r,i_r}} (\varphi_{i,k}(x, u_0, u_1, \ldots, u_r)) =$$

$$\frac{\partial}{\partial u_{0,i_1}}^{\alpha_{1,i_1}} \ldots L^{k_1}_{f_{u_0+u_1}} \cdots \frac{\partial}{\partial u_{0,i_{r-1}}}^{\alpha_{r-1,i_{r-1}}} L^{k_{r-1}}_{f_{u_0+\ldots+u_{r-1}}} \ldots \frac{\partial}{\partial u_{0,i_r}}^{\alpha_{r,i_r}} h_{i,u_0+u_1+\ldots+u_r}.$$

Hence, differentiating the $\varphi_{i,k}(x, u, u_1, \ldots, u_r)$ a certain number of times with respect to the variables $u_{l,j} = (u_l)_j$, $1 \leq j \leq d_u$, $1 \leq l \leq r$ and evaluating at $u_1 = \ldots = u_r = 0$ produces certain functions $\bar{\varphi}_{i,k}(x, u)$, which are generators of $\Xi^\Sigma(u)$.

If $\eta \in \Delta_\Sigma$, $d_X\varphi_{i,k}\,\eta = 0$, which implies that $d_X\bar{\varphi}_{i,k}\,\eta = 0$, which is equivalent to the fact that $\eta \in \bar{\Delta}_\Sigma(u)$.

In the analytic case, we can go further. One has the analytic expansion:

$$\varphi_{i,k}(x, u, u_1, \ldots, u_r) = \sum_\alpha \frac{1}{\alpha!} \Phi_\alpha(x)\, \bar{u}^\alpha, \tag{17}$$

where $\alpha = (\alpha_1, \ldots, \alpha_{rd_u})$, $\alpha! = \alpha_1! \ldots \alpha_{rd_u}!$, $\bar{u}^\alpha = (u_{1,1})^{\alpha_1} \cdots (u_{r,d_u})^{\alpha_{rd_u}}$, and with

$$\Phi_\alpha(x) = (\partial_1)^{\alpha_1} \ldots (\partial_{d_u})^{\alpha_{d_u}} L^{k_1}_{f_u} \cdots L^{k_r}_{f_u} (\partial_1)^{\alpha_{(r-1)d_u+1}} \ldots (\partial_{d_u})^{\alpha_{rd_u}} h_{i,u}(x). \tag{18}$$

This analytic expansion is valid for $u_1, \ldots, u_r$ small enough. Therefore, one has

$$d_X\varphi_{i,k}\eta = \sum_\alpha \frac{1}{\alpha!} d_X \Phi_\alpha(x)\, \eta\, \bar{u}^\alpha, \tag{19}$$

for all $u_1, \ldots, u_r$ small enough, and all tangent vectors $\eta \in T_xX$.

In particular, if $\eta \in \bar{\Delta}_\Sigma(u)$, the right-hand side of (19) vanishes. This shows that $d_X\varphi_{i,k}\eta$ vanishes for $u_1, \ldots, u_r$ small enough. By analyticity and connectedness, $d_X\varphi_{i,k}\eta$ vanishes for all $u, u_1, \ldots, u_r$. This shows that $\eta \in \Delta_\Sigma$. ■

The point of interest, used in Chapter 5, will be that in the analytic controllable case, $\bar{\Delta}_\Sigma(u) = \Delta_\Sigma$ is a regular distribution. This is a consequence of Theorem 5.1 and Theorem 5.2 above.

Exercise 5.3. Find a C^∞ example for which $\bar{\Delta}_\Sigma(u) \neq \Delta_\Sigma$ for some u.

6. Appendix: Weak Controllability

Definition 6.1. A system Σ being given, with state space X, the Lie subalgebra of the Lie algebra of smooth vector fields on X, generated by the vector fields f_u, ($f_u(x) = f(x, u)$), is called the Lie algebra of Σ, and is denoted by $Lie(\Sigma)$.

The Lie algebra $Lie(\Sigma)$ defines an involutive (possibly singular) distribution on X.

Definition 6.2. Let a system Σ be given, with state space X. The system Σ is said weakly controllable, if the Lie algebra $Lie(\Sigma)$ is transitive on X, i.e., $\dim(Lie(\Sigma)(x))$, the dimension of $Lie(\Sigma)$ evaluated at x, as a vector subspace of T_xX, is equal to $n = \dim(X)$, for all $x \in X$.

The following facts are standard, and are used in the book. They come from the classical Frobenius theorem, Chow theorem, and Hermann–Nagano theorem.

1. If a system is weakly controllable, then the **accessibility set** $A_\Sigma(x_0)$ of $x_0 \in X$, i.e., the set of points that can be joined from x_0 by some trajectory of Σ, **in positive time**, has nonempty interior in X, for all $x_0 \in X$.
2. If a system is weakly controllable, then the **orbit** $O_\Sigma(x_0)$ of $x_0 \in X$, i.e. the set of points that can be joined to x_0 by some continuous curve, which is a concatenation of trajectories of Σ **in positive or negative time**, is equal to X, for all $x_0 \in X$.
3. If moreover Σ is analytic, statements (1) and (2) above are "if and only if."
4. If Σ is C^∞, not weakly controllable, but the distribution $Lie(\Sigma)$ on X has constant rank or if Σ is analytic, then $A_\Sigma(x_0)$ has nonempty relative interior in the orbit $O_\Sigma(x_0)$, which is just the integral leaf through x_0 of the distribution $Lie(\Sigma)$.
5. In statements (1), (2), (3), and (4) above, it is possible to restrict to piecewise constant control functions.

3

The Case $d_y \leq d_u$

We will treat only the case $d_y = 1$, $d_u \geq 1$. General results for the case $d_u \geq d_y > 1$ are more difficult to obtain. However, Chapter 8 gives two nontrivial practical examples, where $d_y = d_u = 2$, which are uniformly infinitesimally observable.

In this chapter, with the exception of the first section, we assume that $d_y = 1$, $d_u \geq 1$, and that everything is analytic. We characterize analytic systems that are uniformly infinitesimally observable when restricted to an open dense subset of X. The necessary and sufficient condition is that $\partial_u D(u) = 0$, i.e., **the canonical flag is uniform**. This condition $\partial_u D(u) = 0$ is extremely restrictive and is not preserved by small perturbations of the system.

The analyticity assumption with respect to the x variable is made for purely technical reasons. It can certainly be removed to get similar results (see Exercise 4.6 below).

On the other hand, analyticity with respect to u **is essential.** It is possible to obtain results in the nonanalytic case, but they will be **weaker** in the following sense: to have uniform infinitesimal observability, a certain condition has to hold on an open dense subset of X uniformly in u. In the nonanalytic case, we can only show that this condition has to hold on an open dense subset of $X \times U$. **This is much weaker**. The reason for the "much weaker" result is: To prove the analytic case, we use the permanence properties of projections of semialgebraic or subanalytic sets.

Again d_X (resp. d_U) denotes the differential with respect to the x variables (resp. u variables) only.

1. Relation Between Observability and Infinitesimal Observability

The relation is stated in the following theorem (also valid for $d_y > 1$).

Theorem 1.1.

(i) For any system Σ and any input $\hat{u}$, the set $\theta(\hat{u})$ of states $x \in X$ such that Σ is infinitesimally observable at $(\hat{u}, x)$ is open in X (and could be empty, of course).

(ii) If Σ is observable for an input $\hat{u}$, then $\theta(\hat{u})$ is dense everywhere in X.

(iii) If Σ is infinitesimally observable at $(\hat{u}, x)$, then there exists an open neighborhood V of x such that the restriction $P_{\Sigma,\hat{u}}|V$ is injective (i.e., Σ restricted to V is observable for the input $\hat{u}$).

Proof. Because the output function h depends on u, the output trajectories are measurable functions only and hence, are not uniquely defined pointwise. Some regularization procedure will be needed to palliate this difficulty. For $\hat{u} \in L^{\infty}(U)$, let us define the distribution $D(\hat{u})$: $D(\hat{u})_x = Ker(dP_{\hat{u},x})$. It follows from the definition that Σ is infinitesimally observable at $(\hat{u}, x)$ iff $D(\hat{u})_x = \{0\}$.

For $\xi \in T_x X$, regularize the function $dP_{\hat{u},x}(\xi)(t)$, defined on $[0, e(\hat{u}, x)[$, by setting $\overline{dP_{\hat{u},x}}(\xi)(t) = \int_0^t dP_{\hat{u},x}(\xi)(\tau)d\tau = \omega(x, t)(\xi)$. For each (x, t), $\omega(x, t)$ is a R^{d_y} valued linear form on $T_x X$. The function $(x, t) \in X \times [0, e(\hat{u}, x)[\rightarrow \omega(x, t) \in T_x^* X \otimes R^{d_y}$ is continuous in (x, t), absolutely continuous in t, and analytic in x. It is clear that $D(\hat{u})_x = \bigcap_{t\in[0,e(\hat{u},x)[} Ker(\omega(x, t))$.

There is a finite set of times $t_i \in [0, e(\hat{u}, x)[$, $1 \leq i \leq r$, such that $D(\hat{u})_x = \bigcap_{i=1,\ldots,r} Ker(\omega(x, t_i))$. If Σ is infinitesimally observable at $(\hat{u}, x)$, then $\bigcap_{i=1,\ldots,r} Ker(\omega(x, t_i)) = \{0\}$, and the same happens in a neighborhood of x by the lower semi-continuity of the mapping $x \rightarrow e(\hat{u}, x)$ and the continuity of $\omega(., t_i)$. This shows (i).

Let Y be any open subset of X. Set $c = \sup_{x\in Y} \text{Codim}(D(\hat{u})_x)$. To prove (ii), it is sufficient to prove that $c = n$. Applying the same argument as in the proof of (i), it is clear the set Z of all $x \in Y$ such that $\text{Codim}(D(\hat{u})_x) = c$ is open. Now, consider an open subset W of Z, such that $D(\hat{u})_{|W} = \bigcap_{i=1,\ldots,r} Ker(\omega(x, t_i))$ for some $t_1, \ldots, t_r$. It is clear that $\omega(., t_i) = d_X F_i$, where $F_i : W \rightarrow R^{d_y}$ is the function

$$F_i(x) = \int_0^{t_i} h(\varphi_\tau(\hat{u}, x), \hat{u}(\tau))d\tau.$$

Hence, the restriction $D(\hat{u})|W$ is an integrable distribution, the leaves of which are the connected components of the level manifold of the mappings $F_1, \ldots, F_r$. Let us take a compact connected set K contained in a leaf L of $D(\hat{u})|W$ and containing more than one point. The infimum over K of the function $x \rightarrow e(\hat{u}, x)$ is attained at some x_0. For any time $T \in [0, e(\hat{u}, x_0)[$, there is an open neighborhood V_T of K in W such that $e(\hat{u}, x) > T$ for all $x \in V_T$. Because $\omega(., T)|V_T = d_X F^T$, where $F^T(x) = \int_0^T h(\varphi_\tau(\hat{u}, x), \hat{u}(\tau))d\tau$, and $D(\hat{u})|W \subset Ker(\omega(., T))$, F^T is constant on any connected component of the intersection $L \cap V_T$, in particular on the one containing K. Hence, F^T is constant on K for all $T \in [0, e(\hat{u}, x_0)[$. Differentiating with respect to T shows that $P(\hat{u}, x)(t) = P(\hat{u}, x_0)(t)$ for all $x \in K$, and almost all $t \in [0, e(\hat{u}, x_0)[$. This contradicts the observability assumption. Hence, $c = n$, and this proves (ii). Number (iii) is easy and left to the reader. ■

In the remaining part of the chapter, $d_y = 1$.

2. Normal Form for a Uniform Canonical Flag

We assume that the canonical flag associated to the system Σ is uniform. We will show first that it is equivalent that Σ can be put everywhere locally in a certain normal form, called the **observability canonical form**.

Theorem 2.1. *Σ has a uniform canonical flag if and only if, for all $x_0 \in X$, there is a coordinate neighborhood of x_0, $(V_{x_0}, x^0, \ldots, x^{n-1})$, such that in these coordinates, Σ can be written as follows:*

$$\frac{dx^0}{dt} = f_0(x^0, x^1, u), \ldots, \frac{dx^i}{dt} = f_i(x^0, x^1, \ldots, x^{i+1}, u), \ldots,$$

$$\frac{dx^{n-2}}{dt} = f_{n-2}(x^0, x^1, \ldots, x^{n-1}, u), \frac{dx^{n-1}}{dt} = f_{n-1}(x^0, x^1, \ldots, x^{n-1}, u),$$

$$y = h(x^0, u), \text{ and } \forall (x, u) \in V_{x_0} \times U, \frac{\partial h}{\partial x^0}(x^0, u) \neq 0,$$

$$\frac{\partial f_0}{\partial x^1}(x^0, x^1, u) \neq 0, \ldots, \frac{\partial f_{n-2}}{\partial x^{n-1}}(x^0, \ldots, x^{n-1}, u) \neq 0. \quad (20)$$

Proof. Let us chose coordinates $x^0, \ldots, x^{n-1}$ in a neighborhood V of x_0, such that V is a cube in these coordinates, and

$$D_i(u) = \cap_{j=0,\ldots,i} Ker(dx^j).$$

Then,

$$D_{n-1} = \{0\}, D_{n-2} = Span\left\{\frac{\partial}{\partial x^{n-1}}\right\}, \text{ and for } 0 \leq j \leq n-2,$$

$$L_{\frac{\partial}{\partial x^{n-1}}} L_{f_u}^j h = 0 = L_{f_u} L_{\frac{\partial}{\partial x^{n-1}}} L_{f_u}^{j-1} h + L_{\left[\frac{\partial}{\partial x^{n-1}}, f_u\right]} L_{f_u}^{j-1} h.$$

But,

$$L_{\frac{\partial}{\partial x^{n-1}}} L_{f_u}^{j-1} h = 0,$$

therefore,

$$\left[\frac{\partial}{\partial x^{n-1}}, f_u\right] \in D_{n-3}.$$

An obvious induction shows that

$$\left[\frac{\partial}{\partial x^{k}}, f_u\right] \in D_{k-2},$$

which implies

$$(1)\ \frac{\partial h}{\partial x^i} = 0 \text{ for } i \geq 1,$$

$$(2)\ \frac{\partial f_i}{\partial x^j} = 0 \text{ for } j \geq i+2.$$

Now, it is impossible that $\frac{\partial h}{\partial x^0}(\hat{x}, v) = 0$, or $\frac{\partial f_i}{\partial x^{i+1}}(\hat{x}, v) = 0$ for some i, $0 \leq i \leq n-2$, for $(\hat{x}, v) \in V \times U$, because it implies $\frac{\partial L^i_{f_v} h_v}{\partial x^i}(\hat{x}) = 0$, which contradicts the fact that $D_i(v)(\hat{x}) = Span\{\frac{\partial}{\partial x^{i+1}}, \ldots, \frac{\partial}{\partial x^{n-1}}\}$.

Conversely, if h depends only on u and x^0, if the component f_i depends only on $u, x^0, \ldots, x^{i+1}$, for $1 \leq i \leq n-2$, and if $\frac{\partial h}{\partial x^0} \neq 0$, $\frac{\partial f_i}{\partial x^{i+1}} \neq 0$, for $i = 0, \ldots, n-2$, then it is easy to check that the canonical flag is uniform. ∎

Let us set $h^j_u(x) = h^j(x, u) = L^j_{f_u} h_u(x)$.

Corollary 2.2. *Σ has a uniform canonical flag if and only if, for all $x_0 \in X$ and for all $v \in U$, there exists an open neighborhood $V_{x_0,v}$ of x_0, such that the functions $x^0 = h^0_v|V_{x_0,v}$, $x^1 = h^1_v|V_{x_0,v}$ $\ldots$, $x^{n-1} = h^{n-1}_v|V_{x_0,v}$, form a coordinate system on $V_{x_0,v}$, and on $U \times V_{x_0,v}$, each h^i is a function of u, $x^0, \ldots, x^i$ only, $0 \leq i \leq n-1$.*

Proof. It is clear that the normal form of Theorem 2.1 implies that $x^0 = h^0_v|V_{x_0,v}$, $x^1 = h^1_v|V_{x_0,v}$ $\ldots$, $x^{n-1} = h^{n-1}_v|V_{x_0,v}$, form a coordinate system on a neighborhood of x_0, and each h^i is a function of u, $x^0, \ldots, x^i$ only, $0 \leq i \leq n-1$.

Conversely, if these two statements hold, one computes immediately $D(u)$:

$$D(u) = \left\{ span\left(\frac{\partial}{\partial x^1}, \ldots, \frac{\partial}{\partial x^{n-1}}\right) \supset span\left(\frac{\partial}{\partial x^2}, \ldots, \frac{\partial}{\partial x^{n-1}}\right) \right.$$
$$\left. \supset \ldots \supset span\left(\frac{\partial}{\partial x^{n-1}}\right) \supset \{0\} \right\}.$$

Hence, the canonical flag is uniform. ∎

3. Characterization of Uniform Infinitesimal Observability

The first observation that can be made is the following theorem:

Theorem 3.1. *Assume that Σ is such that its canonical flag is uniform. Then, $\forall x_0 \in X$, there is an open neighborhood V_{x_0} of x_0 such that the restriction $\Sigma_{|V_{x_0}}$ of the system Σ to V_{x_0} is observable and uniformly infinitesimally observable.*

Proof. We apply Theorem 2.1. Take a coordinate neighborhood $(V_{x_0}, x^0, \dots, x^{n-1})$, such that Σ is in the normal form (20). The equation of the first variation for Σ is setting $x(t) = (x^0(t), \dots, x^{n-1}(t))$:

$$\left\{\begin{array}{l} \dot{x}(t) = f(x(t), u(t)), \\ \begin{bmatrix} \dot{\xi}_0 \\ \cdot \\ \cdot \\ \cdot \\ \dot{\xi}_{n-2} \\ \dot{\xi}_{n-1} \end{bmatrix} = \begin{bmatrix} d_X f_0(x^0(t), x^1(t), u(t)) \\ \cdot \\ \cdot \\ \cdot \\ d_X f_{n-2}(x^0(t), \dots, x^{n-1}(t), u(t)) \\ d_X f_{n-1}(x^0(t), \dots, x^{n-1}(t), u(t)) \end{bmatrix} \begin{bmatrix} \xi_0 \\ \cdot \\ \cdot \\ \cdot \\ \xi_{n-2} \\ \xi_{n-1} \end{bmatrix}, \\ \eta = d_X h(x^0(t), u(t))\xi_0. \end{array}\right. \tag{21}$$

We assume that $\Sigma_{|V_{x_0}}$ is not uniformly infinitesimally observable. This means that we can find a trajectory of $\Sigma_{|V_{x_0}}$, and a corresponding trajectory $\xi(t)$ of (21) such that 1) $\xi(0) \neq 0$, 2) $\eta(t) = 0$ a.e..

However, because $d_X h(x^0, u)$ is never 0, this implies that $\xi_0(t) = 0$ everywhere, by continuity. Hence, $\dot{\xi}_0(t) = 0$ almost everywhere. Again, Equation (21) of the first variation, plus the fact that $\partial f_0/\partial x^1$ is never zero, imply that $\xi_1(t) = 0$. An obvious induction shows that $\xi_i(t) = 0$ for $0 \leq i \leq n-1$, which contradicts the fact that $\xi(0) \neq 0$.

It remains to be shown that $\Sigma_{|V_{x_0}}$ is observable. We assume that we can find $u(.)$, and two distinct initial conditions $w(0)$, $z(0)$ such that the corresponding trajectories $w(t)$, $z(t)$ of $\Sigma_{|V_{x_0}}$ produce the same output for almost all $t < T$: $h(w(t), u(t)) = h(z(t), u(t))$ a.e., where T is the minimum of the escape times of both trajectories relative to V_{x_0}.

We can always take V_{x_0} so that its image by $(x^0, \dots, x^{n-1})$ is a cube, and $h(w^0(t), u(t)) - h(z^0(t), u(t)) = 0$ for all t in a subset E of $[0, T[$ of full measure. But, for $t \in E$, it implies that:

$0 = (w^0(t) - z^0(t)) \frac{\partial h}{\partial x^0}(c(t), u(t))$, for some $c(t)$, $w^0(t) \leq c(t) \leq z^0(t)$ if $w^0(t) \leq z^0(t)$.

Because $\partial h/\partial x^0$ never vanishes, we get that $w^0(t) = z^0(t)$ almost everywhere, and hence everywhere because $w^0(t)$ and $z^0(t)$ are absolutely continuous functions.

An induction similar to the one used in the first part of the proof shows that, $w(t) = z(t)$ for all $t \in [0, T[$. This contradicts the fact that $w(0) \neq z(0)$. ■

Remark 3.1. We have proved a bit more than the statement of the theorem: we have proved that if Σ has the normal form (20) on a **convex** subset C of R^n, then the restriction $\Sigma_{|C}$ is observable and uniformly infinitesimally observable.

The main result in this chapter is that, conversely, **the uniformity condition on the canonical flag is a necessary condition** for uniform infinitesimal observability, at least on an open dense subset of X.

Let us point out the following fact about this result: it is true "almost everywhere" with respect to X, but it is **global** with respect to U. This is the hard part to prove. If one is interested with a result true almost everywhere with respect to both x and u, **the proof is much easier**.

Before proceeding, let us make the following standing assumptions: **either,**

(H_1) $U = I^{d_u}$, $I \subset R$ is a compact interval, and the system is analytic, **or,**

(H_2) $U = R^{d_u}$, f and h are algebraic with respect to u.

Let $\tilde{M}$ be the subset of $U \times X$:

$$\tilde{M} = \left\{(u, x) | d_X h_u^0(x) \wedge \ldots \wedge d_X h_u^{n-1}(x) = 0\right\}.$$

Let M be its projection on X. Then:

Theorem 3.2. *Assume either (H_1) or (H_2) and that Σ is uniformly infinitesimally observable. Then:*

1. *The set M is a subanalytic (resp. semianalytic in case of (H_2)) set* ***of codimension at least*** 1*. In the case (H_1), M is closed. In any case, denote by $\bar{M}$ its closure,*
2. *The restriction $\Sigma_{|X \backslash \bar{M}}$ of Σ to $X \backslash \bar{M}$ has a uniform canonical flag.*

This theorem will be proved in Section 5 of this chapter. Before this, let us give some comments and examples, and state some complementary results.

4. Complements

4.1. Exercises

Exercise 4.1. Let Σ be a system with uniform canonical flag. Show that Σ is strongly differentially observable of order n, and hence has the phase variable property of order n, when restricted to sufficiently small open subsets of X.

Exercise 4.2. Show that the (uncontrolled) system on R^2:

$$\dot{x}_1 = x_2, \dot{x}_2 = 0, \; y = (x_1)^3,$$

is observable on R^2, but not infinitesimally observable at $x_0 = 0$.

Exercise 4.3. The output function does not depend on u, and $n = \dim(X) = 2$. Assume that we work in the class of systems such that h does not depend on u, $u \in I^{d_u}$, I compact. Fix $x^0 \in X$.

1. Show that the property that Σ has a uniform canonical flag in a neighborhood of x^0 is stable under C^2-small perturbations of Σ.
2. Show that, if $n > 2$, this is not true.

Exercise 4.4. In the class of control affine systems (i.e., $\dot{x} = f(x) + u\, g(x)$, $y = h(x)$, $U = R$), show that the result 1 of Exercise 4.3 is false.

Exercise 4.5. Show that the system on $X = R$:

$$\dot{x} = 1,$$
$$y = \frac{1}{2}\left(x - \frac{\sin\left(2x(1+u^2)^{\frac{1}{2}}\right)}{2(1+u^2)^{\frac{1}{2}}}\right) + x \sin^2 u, \; u \in R$$

is uniformly infinitesimally observable, but Theorem 3.2 is false (U is **not compact**).

4.2. Control Affine Systems

We consider the control affine analytic systems (with single control, to simplify):

$$(\Sigma_A) \begin{cases} \dot{x} = f(x) + u\, g(x), \\ y = h(x), \; u \in R. \end{cases} \tag{22}$$

In that case, there is a stronger statement than Theorem 3.2, which is much easier to prove. Consider the mapping $\Phi : X \to R^n$, $\Phi(x) = (h(x), L_f h(x), \ldots, L_f^{n-1} h(x))$. The set of points $x \in X$ at which Φ is not a local diffeomorphism, i.e., $d_X h(x) \wedge d_X L_f h(x) \wedge \ldots \wedge d_X L_f^{n-1} h(x) = 0$ is an analytic subset, closed in X, denoted by M.

Theorem 4.1.

1. *If Σ_A is* **observable**, *then M has codimension* 1 *at least, and, on each open subset $Y \subset X \backslash M$ such that the restriction $\Phi_{|Y}$ is a diffeomorphism, $\Phi_{|Y}$ maps $\Sigma_{A|Y}$ into a system $\bar{\Sigma}_A$ of the form*

$$(\bar{\Sigma}_A)\ \dot{x} = \begin{pmatrix} \dot{x}_1 \\ \dot{x}_2 \\ \cdot \\ \cdot \\ \cdot \\ \dot{x}_{n-1} \\ \dot{x}_n \end{pmatrix} = \begin{pmatrix} x_2 \\ x_3 \\ \cdot \\ \cdot \\ \cdot \\ x_n \\ \varphi(x) \end{pmatrix} + u \begin{pmatrix} g_1(x_1) \\ g_2(x_1, x_2) \\ \cdot \\ \cdot \\ \cdot \\ g_{n-1}(x_1, \ldots, x_{n-1}) \\ g_n(x) \end{pmatrix}. \quad (23)$$

$$y = x_1.$$

2. *Conversely, if Ω is an open subset of R^n on which the system Σ has the form $\bar{\Sigma}_A$, then the restriction $\Sigma_{|\Omega}$ is observable.*

The proof is simple, and contains the basic idea for the proof of Theorem 3.2.

Proof of Theorem 4.1. The second part is very simple: Assume that two trajectories corresponding to $u(t)$, say $x(t)$ and $\tilde{x}(t)$, produce the same output $x_1(t) = \tilde{x}_1(t)$. Then, $\frac{d(x_1(t)-\tilde{x}_1(t))}{dt} = 0$ almost everywhere. It follows from (23) that almost everywhere, $0 = (x_2(t) - \tilde{x}_2(t)) + u(t)(g_1(x_1(t) - g_1(\tilde{x}_1(t))$. Hence, $0 = (x_2(t) - \tilde{x}_2(t))$ almost everywhere, and by continuity, $0 = (x_2(t) - \tilde{x}_2(t))$ everywhere (i.e., for all $t \in [0, \min(e(u, x(0)), e(u, \tilde{x}(0))[$). An obvious induction shows that $x(t) = \tilde{x}(t)$ everywhere, and in particular $x(0) = \tilde{x}(0)$.

For the first part, it is clear that the set M is analytic. Assume that its codimension is zero. Then $M = X$ and $d_{n-1}(x) = dh(x) \wedge dL_f h(x) \wedge \ldots \wedge dL_f^{n-1} h(x)$ is identically zero. Let $k \leq n-1$ be the first index such that $d_k \equiv 0$. We cannot have $k = 0$ because it implies that h is a constant function, and then the system Σ is not observable. We can take a small open coordinate system $(x^1, \ldots, x^n)$ with first k coordinates $(h(x), L_f h(x), \ldots, L_f^{k-1} h(x))$. Then, in these coordinates, $L_f^n h$ is a function of $(x^1, \ldots, x^k)$ for all n: $d_k(x) = 0 = dx^1 \wedge \ldots \wedge dx^k \wedge dL_f^k h(x)$. Hence, $L_f^k h(x) = \varphi(x^1, \ldots, x^k)$ for a certain smooth function φ. In the same way, $L_f^{k+1} h(x) = L_f \varphi(x) = \frac{\partial \varphi}{\partial x^1} x^2 + \ldots + \frac{\partial \varphi}{\partial x^{k-1}} x^k + \frac{\partial \varphi}{\partial x^k} \varphi = \psi(x^1, \ldots, x^k)$ for a certain smooth function ψ. By induction, we see that all the functions $L_f^n h(x)$ are functions of

$(x^1, \ldots, x^k)$ only. Hence they do not separate the points, and by (14) in Chapter 2, Σ is not observable for the control function $u(.) \equiv 0$.

Now, let Y be an open subset of X such that $\Phi_{|Y}$ is a diffeomorphism. On Y, $(h(x), L_f h(x), \ldots, L_f^{n-1} h(x))$ is a coordinate system. Let us show that g has the required form, in these coordinates (it is already clear that f and h have the required form). Assume that the first component g_1 of g depends on x^i, for some $i > 1$. Let us consider two initial conditions $x_1, x_2 \in Y$, such that (a) $x_1^1 = x_2^1$, and (b) $g_1(x_1) \neq g_1(x_2)$. This is possible because of the assumption that g_1, does not depend on x^1 only. We will construct a control function $u(t)$, such that these two initial conditions together with the control $u(t)$ produce the same output function. For this, let us consider the product system $\Sigma \times \Sigma$ of Σ by itself:

$$\begin{aligned} \dot{x}_1^* &= f(x_1^*) + u\, g(x_1^*), \\ \dot{x}_2^* &= f(x_2^*) + u\, g(x_2^*), \\ y &= h(x_1^*) - h(x_2^*). \end{aligned} \tag{24}$$

The state space of $\Sigma \times \Sigma$ is $X \times X$. Let us substitute in $\Sigma \times \Sigma$ the following function $\hat{u}$ for u:

$$\hat{u}(x_1^*, x_2^*) = -\frac{L_f h(x_1^*) - L_f h(x_2^*)}{L_g h(x_1^*) - L_g h(x_2^*)}. \tag{25}$$

The resulting differential system is well defined in a neighborhood of the initial condition (x_1, x_2) because the function $\hat{u}$ is smooth in a neighborhood of (x_1, x_2): The condition $g_1(x_1) \neq g_1(x_2)$ is equivalent to $L_g h(x_1) - L_g h(x_2) \neq 0$. Call $(x_1^*(t), x_2^*(t))$ the trajectory of this differential system, starting at (x_1, x_2).

It is easy to check that, for the control function $u(t) = \hat{u}(x_1^*(t), x_2^*(t))$, the output of $\Sigma \times \Sigma$ is identically zero (for small times). This shows that Σ is not observable for $u(t)$. Similarly, an induction shows that g_i depends only on $(x^1, \ldots, x^i)$. ∎

Exercise 4.6. Give a statement and a proof of Theorem 4.1 in the C^∞ case.

4.3. Bilinear Systems (Single Output)

Bilinear systems are systems on $X = R^n$, that are control affine **and** state affine:

$$(\mathcal{B}) \begin{cases} \dot{x} = Ax + u\,(Bx + b), \\ y = Cx, \end{cases} \tag{26}$$

where $A : R^n \to R^n$, $B : R^n \to R^n$ are linear, $b \in R^n$, $C \in (R^n)^*$. For these bilinear systems, the previous result, Theorem 4.1, can be made much stronger.

Exercise 4.7. Show the following theorem:

Theorem 4.2. *The single-output bilinear system ($\mathcal{B}$) is* ***observable*** *if and only if it has the following form in the (linear) coordinate system* $x^* = (Cx, CAx, \ldots, CA^{n-1}x)$: $\dot{x}^* = \bar{A}x^* + u(\bar{B}x^* + \bar{b})$, $y = \bar{C}x$, *where* $\bar{C} = (1, 0, \ldots., 0)$, $\bar{B}$ *is lower triangular, and* $\bar{A}$ *is a companion matrix:*

$$\bar{A} = \begin{pmatrix} 0, 1, 0, \ldots\ldots\ldots, 0 \\ 0, 0, 1, 0, \ldots\ldots\ldots, 0 \\ \vdots \\ 0, \ldots\ldots\ldots\ldots, 0, 1 \\ a_1, \ldots\ldots\ldots\ldots, a_n \end{pmatrix}. \tag{27}$$

The bilinear systems (single output or not) play a very special role from the point of view of the observability property: the *initial − state → output − trajectory* mapping is an affine mapping. In fact, the control function being known, they are just linear time-dependent systems. Therefore, for instance, the observer problem can be solved just by using the linear theory.

A very important result is stated in the following exercise.

Exercise 4.8. (Fliess–Kupka theorem in the analytic case)

1. Define a reasonable notion of an immersion of a system into another one.
2. Show that a control affine analytic system can be immersed into a bilinear one if and only if its observation space Θ^{Σ} is finite-dimensional. (See Chapter 2, Section 5, for the definition of the observation space.)

This result ((2) in Exercise 4.8) is not very difficult to prove. The original result, in paper [15], is a similar theorem in the C^{∞} case, the proof of which is not that easy.

5. Proof of Theorem 3.2

We will need the following lemma:

Lemma 5.1. *Let Y be a* ***connected*** *analytic manifold, $Z \subset R^d$ be an open neighborhood of $x_0 \in R^d$, and let $f^0, \ldots, f^n : Y \times Z \to R$ be $n+1$ analytic functions such that:*

1. $d_Z f^0 \wedge \ldots \wedge d_Z f^n \wedge d_Y d_Z f^n = 0$ *on* $Y \times Z$,
2. *There exists* $y_0 \in Y$ *such that:* $f^0_{y_0}, \ldots, f^n_{y_0} : Z \to R$ *are independent,*

3. There exist analytic functions $h^i : Y \times \Theta_i \to R$, Θ_i open in R^n, $0 \leq i \leq n-1$, such that: $f^i(y, z) = h^i(y, f^0_{y_0}, \ldots, f^{n-1}_{y_0}(z))$ for all $(y, z) \in Y \times Z$.

Then, there exists a subneighborhood $Z' \subset Z$ of x_0, and a function $h^n : Y \times \Theta_n \to R$, $\Theta_n \subset R^{n+1}$, Θ_n open, such that for all $(y, z) \in Y \times Z'$, $f^n(y, z) = h^n(y, f^0_{y_0}, \ldots, f^n_{y_0}(z))$.

Proof. Let $Z' \subset Z$ be an open subneighborhood of x_0, $Y_0 \subset Y$ be an open neighborhood of y_0 in Y, and $\varphi = (x^0, \ldots, x^d) : Z \to R^d$, a global coordinate system on Z' such that $x^i = f^i_{y_0}$, $0 \leq i \leq n$, and,

(i) $\varphi(Z')$ is a convex subset of R^d,
(ii) $f^0_y, \ldots, f^n_y, x^{n+1}, \ldots, x^d : Z' \to R$, is a coordinate system on Z' for all $y \in Y_0$.

By restricting Y_0, we can assume that it carries a coordinate system $\tilde{y} = y^1, \ldots, y^m : Y_0 \to R$, such that $\tilde{y}(Y_0)$ is convex and $\tilde{y}(y_0) = 0$. Then, $d_Z f^0 \wedge \ldots \wedge d_Z f^n \wedge d_Y d_Z f^n = 0$ on $Y \times Z$ implies:

$$d_Z f^0 \wedge \ldots \wedge d_Z f^n \wedge d_Z\left(\frac{\partial f^n}{\partial y^i}\right) = 0 \text{ on } Y_0 \times Z \text{ for all } 1 \leq i \leq m.$$

This in turn implies that

$$f^n(ty, z) = \int_0^t \sum_{k=1}^m h^n_k(sy, x^0(z), \ldots, x^{n-1}(z), f^n(sy, z))y^k ds + f^n(y_0, z), \tag{28}$$

on $Y_0 \times Z'$, for some functions $h^n_k : Y_0 \times \bar{\Theta}$, $\bar{\Theta} \subset R^{n+1}$, $\bar{\Theta}$ open.

Now, there exists a neighborhood $Y'_0 \subset Y_0$ of y_0 such that Equation (28) has a unique solution (given $f^n(y_0, z)$) defined on $Y'_0 \times Z'$ and for all $t \in [0, 1]$. This solution is an analytic function of $t, y, x^0, \ldots, x^{n-1}, f^n_{y_0}$. If we take $t = 1$, we obtain

$$f^n(y, z) = H^n\left(y, x^0(z), \ldots, x^{n-1}(z), f^n_{y_0}\right),$$

on $Y'_0 \times Z'$, or

$$f^n_y = H^n(y, x^0, \ldots, x^n),$$

on $Y'_0 \times Z'$.

However, because $x^0, \ldots, x^d : Z \to R$, is a coordinate system over Z', we have $f^n = G(y, x^0, \ldots, x^d)$ on $Y \times Z'$. On the open subset $Y'_0 \times Z'$,

$$\frac{\partial G}{\partial x^j} = \frac{\partial H^n}{\partial x^j} = 0, n + 1 \leq j \leq d.$$

Hence, $\frac{\partial G}{\partial x^j} = 0$ on $Y \times Z'$, for $n + 1 \leq j \leq d$.

Again, by the convexity of Z',

$$G(y, x^0, \ldots, x^d) =$$

$$\int_0^1 \left(\frac{\partial G\left(y, x^0, \ldots, x^n, tx^{n+1} + (1-t)x_0^{n+1}, \ldots, tx^d + (1-t)x_0^d\right)}{\partial x^{n+1}} \left(x^{n+1} - x_0^{n+1}\right) \right.$$

$$\left. + \ldots + \frac{\partial G\left(y, x^0, \ldots, x^n, tx^{n+1} + (1-t)x_0^{n+1}, \ldots, tx^d + (1-t)x_0^d\right)}{\partial x^d} \left(x^d - x_0^d\right) \right) dt$$

$$+ G\left(y, x^0, \ldots, x^n, x_0^{n+1}, \ldots, x_0^d\right).$$

Hence, $f^n = G(y, x^0, \ldots, x^n)$ on $Y \times Z'$. ■

We will prove, by induction on N, the following statement (A_N):

Let M_N be the projection on X of the semi-analytic (resp. analytic, partially algebraic) subset $\tilde{M}_N = \{(u, x) | d_X h_u^0(x) \wedge \ldots \wedge d_X h_u^N(x) = 0\}$:

- M_N is a subanalytic (respectively semi-analytic) subset of X of codimension ≥ 1, and
- For any $a \in X \backslash \bar{M}_N$, and any $v \in U$, there exists an open neighborhood V of a such that the restriction of h^i to $U \times V$ is a function of u and of the restrictions $h_{v|V}^0, \ldots, h_{v|V}^i$ of $h_v^0, \ldots, h_v^i$ to V only, for all i, $0 \leq i \leq N$ (this second property we denote by (P_N)).

It is clear, by Corollary 2.2 that the property (A_{n-1}) implies Theorem 3.2. In fact, because all the functions $h_v^0, \ldots, h_v^{n-1}$ are independent on V, there exists a (could be smaller) neighborhood V_a of a, such that $h_v^0, \ldots, h_v^{n-1}$ form an analytic coordinate system on V_a. Also, $M_n = M$, $M_{n-1} \supset \ldots \supset M_1 \supset M_0$.

Assume that we have proved that $A_0, \ldots, A_{N-1}$, and let us prove A_N. This will be done in four steps. In order to prove steps 1, 2, and 4, we will construct feedback laws contradicting infinitesimal observability. In step 1, this feedback will be a constant control. In step 2, it is a general feedback for the lift of our system on the tangent bundle (i.e., a feedback depending on ξ, with the notations of Section 1, Chapter 2), and in step 4, it will be a feedback depending on x only.

Let Z_N be the set of all $x \in X$ such that $dh_u^0(x) \wedge \ldots \wedge dh_u^N(x) = 0$ for all $u \in U$. Because $Z_N = \cap_{u \in U} Z_N(u)$, where $Z_N(u) = \{x \in X \mid dh_u^0(x) \wedge \ldots \wedge dh_u^N(x) = 0\}$, is an analytic subset of X, it follows that Z_N also is analytic ([43], Corollary 2, p. 100).

Step 1. We Claim that the Codimension of Z_N is at Least 1

Were it otherwise, Z_N would contain an open set Ω. Then, $\Omega \backslash \bar{M}_{N-1}$ is also open and nonempty. Because for any $u \in U$, $dh_u^0(x) \wedge \ldots \wedge dh_u^N(x) = 0$ on $\Omega \backslash \bar{M}_{N-1}$, any point $a \in \Omega \backslash \bar{M}_{N-1}$ has, for a given $v \in U$, an open neighborhood W in $\Omega \backslash \bar{M}_{N-1}$, such that h_v^N is a function of $h_v^0, \ldots, h_v^{N-1}$ in W.

If $\xi : [0, e[\to TW$ is any trajectory of $T\Sigma$ corresponding to the constant control $u(t) = v$, and such that

$$\xi(0) \in T_a X, \xi(0) \neq 0, dh_v^0(\xi(0)) = \ldots = dh_v^{N-1}(\xi(0)) = 0,$$

then we will see that $dh_v^0(\xi(t)) = \ldots = dh_v^{N-1}(\xi(t)) = 0$ for all $t \in [0, e[$. In fact, $\frac{d}{dt}(dh_v^i(\xi(t)) = d(L_{f_v}(h_v^i))(\xi) = dh_v^{i+1}(\xi(t))$ for all i, $0 \leq i \leq N-1$.

Because $dh_v^N(\xi)$ is a linear combination of $dh_v^0(\xi), \ldots, dh_v^{N-1}(\xi)$, it follows from the uniqueness part of the Cauchy theorem for ordinary differential equations that $dh_v^0(\xi(t)) = \ldots = dh_v^{N-1}(\xi(t)) = 0$. Thus, Σ is not uniformly infinitesimally observable, which is a contradiction.

Step 2. On $TU \times T(X \backslash \bar{M}_{N-1}), d_X h^0 \wedge \ldots \wedge d_X h^N \wedge d_U d_X h^N = 0$

If $x^1, \ldots, x^n$ is a coordinate system on some open subset X' of X, and $u^1, \ldots, u^{d_u}$ is a coordinate system on some open subset U' of U, then

$$d_X h^i = \sum_{j=1}^{n} \frac{\partial h^i}{\partial x^j}(u, x) dx^j, \; d_U d_X h^N = \sum_{k=1}^{d_u} \sum_{j=1}^{n} \frac{\partial^2 h^N}{\partial u^k \partial x^j} du^k \wedge dx^j. \quad (29)$$

Assume that the assertion of step 2 is not true. Then there exists a pair $(u_0, \xi_0) \in Int(U) \times T_a(X \backslash \bar{M}_{N-1})$ such that

$$d_X h_{u_0}^i(\xi_0) = 0 \text{ for } 0 \leq i \leq N, \quad (30)$$

but $d_U d_X h^N(., \xi_0)$ is not identically zero on $T_{u_0} U$.

We will construct an analytic feedback $\bar{u} : W \to U$, W an open neighborhood of ξ_0 in $T(X \backslash \bar{M}_{N-1})$, $\bar{u}(\xi_0) = u_0$, and a solution $\hat{\xi} : [0, e[\to W$ of the feedback system

$$\frac{d\xi}{dt} = T_X f(\xi, \bar{u}(\xi)), \; \hat{\xi}(0) = \xi_0,$$

such that $d_X h_{\hat{u}(t)}^i(\hat{\xi}(t)) = 0$ for all $t \in [0, e[$ and $0 \leq i \leq N-1$, where $\hat{u}(t) = \bar{u}(\hat{\xi}(t))$. This contradicts the infinitesimal observability assumption.

The feedback $\bar{u}$ is the solution of $d_X h_{\bar{u}(\xi)}^N(\xi) = 0$, for all $\xi \in W$, $\bar{u}(\xi_0) = u_0$, obtained by the implicit function theorem. We have

$$\frac{d}{dt}(d_X h_{\hat{u}(t)}^0(\hat{\xi}(t))) = d_X h_{\hat{u}(t)}^1(\hat{\xi}(t)) + d_U d_X h^0\left(\frac{d\hat{u}(t)}{dt}, \hat{\xi}(t)\right),$$

$$\frac{d}{dt}(d_X h_{\hat{u}(t)}^i(\hat{\xi}(t))) = d_X h_{\hat{u}(t)}^{i+1}(\hat{\xi}(t)) + d_U d_X h^i\left(\frac{d\hat{u}(t)}{dt}, \hat{\xi}(t)\right), 1 \leq i \leq N-2,$$

$$\frac{d}{dt}(d_X h_{\hat{u}(t)}^{N-1}(\hat{\xi}(t))) = d_U d_X h^{N-1}\left(\frac{d\hat{u}(t)}{dt}, \hat{\xi}(t)\right), \quad (31)$$

because by construction,

$$d_X h^N_{\bar{u}(\hat{\xi}(t))}(\hat{\xi}(t)) = d_X h^N_{\hat{u}(t)}(\hat{\xi}(t)) = 0. \tag{32}$$

We apply the induction assumption to $a = \Pi(\xi_0)$ and u_0. We call V_a the corresponding neighborhood. Restrict both W and V_a so that $\Pi(W) \subset V_a$. For any $v \in \bar{u}(W)$, and any $0 \leq i \leq N-1$, h^i is a function of u and $h^0_v, \ldots, h^i_v$ only on $U \times V_a$. It follows that

$$d_U d_X h^i \left(\frac{d\hat{u}(t)}{dt}, \hat{\xi}(t) \right)$$

is a linear combination of $d_X h^0_{\hat{u}(t)}(\hat{\xi}(t)), \ldots\ldots, d_X h^i_{\hat{u}(t)}(\hat{\xi}(t))$, for all $0 \leq i \leq N-1$ and all t.

The system (31) becomes a linear time-dependent system in the functions $d_X h^i_{\hat{u}(t)}(\hat{\xi}(t))$. At $t = 0$, $d_X h^i_{\hat{u}(t)}(\hat{\xi}(t))_{|t=0} = d_X h^i_{u_0}(\xi_0) = 0$, by (30). By Cauchy's uniqueness theorem, we get that $d_X h^i_{\hat{u}(t)}(\hat{\xi}(t)) = 0$ for all t. This contradicts the fact that Σ is uniformly infinitesimally observable.

Step 3. Proof of (P_N)

Take any point a in $X \backslash (Z_N \cup \bar{M}_{N-1})$. There exists a $v \in Int(U)$ such that (v, a) is not in $\tilde{M}_N$. We know that $d_X h^0 \wedge \ldots \wedge d_X h^N \wedge d_U d_X h^N = 0$ every where on $X \backslash \bar{M}_{N-1}$. Now, we can apply (A_{N-1}) to a and v. Because (v, a) is not in $\tilde{M}_N$, $d_X h^0_v(a) \wedge \ldots \wedge d_X h^N_v(a) \neq 0$. Restricting the neighborhood V_a given by (A_{N-1}), we can assume that the set $\{h^0_v, \ldots, h^N_v\}$ can be extended to a coordinate system in V_a, meeting the assumptions of Lemma 5.1. Applying this lemma to $Y = U$, $Z = V_a$, $f^i = h^i$, we get that h^N is a function of u and $h^0_v, \ldots, h^N_v$ only in $U \times V_a$.

Step 4. Proof of the Fact that M_N has Codimension 1

We chose a and V_a as in the step 3. For simplicity let us denote the restrictions $h^0_v | V_a, \ldots, h^N_v | V_a$ by $x^0, \ldots, x^N$. Then $h^i = H^i(u, x^0, \ldots, x^i)$ for $0 \leq i \leq N$ in $U \times V_a$. We have

$$d_X h^0 \wedge \ldots \wedge d_X h^N = \frac{\partial H^0}{\partial x^0} \frac{\partial H^1}{\partial x^1} \cdots \frac{\partial H^N}{\partial x^N} dx^0 \wedge \ldots \wedge dx^N,$$

$$d_X h^0 \wedge \ldots \wedge d_X h^{N-1} = \frac{\partial H^0}{\partial x^0} \frac{\partial H^1}{\partial x^1} \cdots \frac{\partial H^{N-1}}{\partial x^{N-1}} dx^0 \wedge \ldots \wedge dx^{N-1}.$$

Because $V_a \cap \bar{M}_{N-1} = \varnothing$, $\frac{\partial H^0}{\partial x^0}, \frac{\partial H^1}{\partial x^1}, \ldots, \frac{\partial H^{N-1}}{\partial x^{N-1}}$ are all everywhere nonzero in $U \times V_a$. Because $\tilde{M}_N = \{(u, x) \in U \times X \mid dh^0_u(x) \wedge \ldots \wedge dh^N_u(x) = 0\}$, we see that

$$\tilde{M}_N \cap (U \times V_a) = \left\{(u, x) \in U \times V_a \mid \frac{\partial H^N}{\partial x^N}(u, x) = 0\right\}.$$

What remains to be proven is that M_N has an empty interior. Otherwise, M_N would contain an open set Θ, and $\Theta\backslash(Z_N \cup \bar{M}_{N-1})$ would be a nonempty open set. Take a point $a \in \Theta\backslash(Z_N \cup \bar{M}_{N-1})$. Apply the considerations just developed to a and restrict the neighborhood V_a, we have constructed $V_a \cap (\Theta\backslash(Z_N \cup \bar{M}_{N-1}))$.

Denote by $P : \tilde{M}_N \cap (U \times V_a) \to V_a$ the restriction of the projection $U \times V_a \to V_a$ to $\tilde{M}_N$. Because P is surjective, Sard's theorem and the implicit function theorem show that there is an open subset W of V_a, and an analytic mapping $\bar{u} : W \to U$, such that

$$\frac{\partial H^N}{\partial x^N}(\bar{u}(x), x) = 0 \text{ for all } x \in W.$$

The same reasoning as before will show that Σ cannot be uniformly infinitesimally observable. Let $\hat{\xi} : [0, e[\to TW$ be any maximal (for positive times) solution of the feedback system

$$\frac{d\hat{\xi}}{dt} = T_X f(\bar{u}(\Pi(\hat{\xi})), \hat{\xi}) \text{ in } TW, \text{ such that } \hat{\xi}(0) \neq 0,$$

but

$$d_X h^i_{\bar{u}(x_0)}(\hat{\xi}_0) = 0 \text{ for } 0 \leq i \leq N-1,\ x_0 = \Pi(\hat{\xi}_0).$$

As before, we have

$$\frac{d}{dt} d_X h^i_{\hat{u}}(\hat{\xi}) = d_X h^{i+1}_{\hat{u}}(\hat{\xi}) + d_U d_X h^i\left(\frac{d\hat{u}}{dt}, \hat{\xi}\right),\ 0 \leq i \leq N-1,$$

where $\hat{u}(t) = \bar{u}(\Pi(\hat{\xi}(t)))$, $\hat{u} : [0, e[\to U$.

Because $\frac{\partial H^i}{\partial x^i} \neq 0$ in $U \times V_a$, $0 \leq i \leq N-1$, $d_U d_X h^i(\frac{d\hat{u}}{dt}, \hat{\xi})$ is a linear combination of $d_X h^0_{\hat{u}}(\hat{\xi}), \ldots, d_X h^i_{\hat{u}}(\hat{\xi})$.

Also,

$$d_X h^N_{\hat{u}}(\hat{\xi}) = \sum_{j=0}^{N} \frac{\partial H^N}{\partial x^j}(\hat{u}, \Pi(\hat{\xi})) dx^j(\hat{\xi}).$$

However,

$$\frac{\partial H^N}{\partial x^N}(\hat{u}, \Pi(\hat{\xi})) = \frac{\partial H^N}{\partial x^N}(\bar{u}(\Pi(\hat{\xi})), \Pi(\hat{\xi})) = 0.$$

So $d_X h_{\hat{u}}^N(\hat{\xi})$ is again a linear combination of $d_X h_{\hat{u}}^0(\hat{\xi}), \ldots, d_X h_{\hat{u}}^{N-1}(\hat{\xi})$. Again we can apply Cauchy's uniqueness theorem to get a contradiction. Thus, M_N, and hence $\bar{M}_N$ have codimension 1, because the interior of M_N is empty. This ends the proof of Theorem 3.2. ■

4

The Case $d_y > d_u$

We refer to the notations of Chapter 2, Sections 3 and 4. The main purpose in this chapter is to show that, in the case of $d_y > d_u$, the picture is completely reversed: Roughly speaking, the observability becomes a generic property. More precisely, the **strong differential observability property of order** $2n + 1$ (in the sense of the Definition 4.2, Chapter 2) is generic (in the Baire sense only: It is an open problem to prove that the set of strongly differentially observable systems contains an open dense set of systems. If we were able to show this openness property, some deep technical complications could be avoided in the proof of the other results: in particular, the "complexification step" below).

Another very important result is the following. **Observability (for all L^∞ inputs) is a dense property,** that is, any system can be approximated by an observable one.

In the case where X is compact, strong differential observability means that $S\Phi_N^\Sigma$ is an embedding, or that $\Phi_N^\Sigma(., \underline{u}_N)$ is an embedding for all $\underline{u}_N$. Therefore, the set of systems, such that $S\Phi_N^\Sigma$ is an embedding, is generic, if $N \geq 2n + 1$.

Of course, there is no chance to prove such a general result if X is not compact:

Exercise 0.1. Show that, even among ordinary smooth mappings between finite dimensional manifolds X, Y, embeddings may not be dense, whatever the dimension $N = \dim(Y)$ with respect to $n = \dim(X)$.

(For hints, see [26, p. 54].)

The reason for the fact stated in Exercise 0.1 is that embeddings are proper mappings. For our practical purposes (synthesis of observers and output stabilization), we don't need that the fundamental mapping $S\Phi_N^\Sigma$ be proper. It is sufficient for it to be an injective immersion. Hence, all the genericity results we prove are true for a noncompact X and for the Whitney topology. But for the sake of simplicity, we shall assume in this chapter that X **is a compact manifold**, and leave all generalizations to the reader as exercises.

Also, we will assume that X is an **analytic manifold**. One should be conscious of the fact that this is not a restriction: Any C^∞ manifold possesses a compatible C^ω structure (see [26, p. 66]).

Again, in this chapter, $U = I^{d_u}$, where I is a closed bounded interval. Because we make an extensive use of subanalytic sets and their properties, this compactness assumption cannot be relaxed.

1. Definitions and Notations

The systems under consideration are of the form

$$(\Sigma)\ \frac{dx}{dt} = f\big(x, u^{(0)}\big)\ ;\ y = h\big(x, u^{(0)}\big) \tag{33}$$

or

$$(\Sigma)\ \frac{dx}{dt} = f\big(x, u^{(0)}\big)\ ;\ y = h(x), \tag{34}$$

in order to take into account the more practical cases in which the output function h does not depend on u: The proofs of the genericity results in that case are not different from the proofs in the general case, where h depends on u, but these results do not follow from the results in the general case.

In agreement with the notations introduced in Section 4 of Chapter 2, we shall use the notation $u^{(0)}$ for the control variable.

For technical reasons, we will have to handle these two classes of systems in the C^r case, **and also some other classes of systems**. Let us explain this below.

We will assume that f and h are at least C^r w.r.t. $(x, u^{(0)})$, **for r large enough**. We shall endow the set of systems with the topology of C^r uniform convergence over $X \times U$. The set of systems with this topology will be denoted by S^r. In the particular case where h does not depend on u, it will be denoted by $S^{0,r}$. Then, $S^r = F^r \times H^r$, $S^{0,r} = F^r \times H^{0,r}$, where F^r denotes the set of u-parametrized vector fields f over X, that are C^r with respect to both x and u. Also, H^r denotes the set of C^r maps h: $X \times U \to R^{d_y}$, and $H^{0,r}$ denotes the set of C^r maps h: $X \to R^{d_y}$. The spaces F^r, H^r, and $H^{0,r}$ will also be endowed with the C^r topology.

The spaces H^r, $H^{0,r}$, and F^r are the Banach spaces of C^r sections of the following bundles B_H, B_{H^0}, B_F over $X \times U$, X and $X \times U$ respectively: $B_H = X \times U \times R^{d_y}$, $B_{H^0} = X \times R^{d_y}$, $B_F = TX \times U$. The spaces $S^{0,r}$, $H^{0,r}$ are closed subspaces of the Banach spaces S^r, H^r, respectively.

The bundles of k-jets of C^r sections of B_F, B_H, B_{H^0} are denoted by J^kF, J^kH, J^kH^0, respectively.

The symbol $\times_X$ will mean the standard fiber-product of bundles over X. For bundles $\Pi_E: E \to X \times U$ and $\Pi_F: F \to X$, we set $E\underline{\times_X}F = \{(e, f) \mid pr_1 \circ \Pi_E(e) = \Pi_F(f)\}$, with $pr_1 : X \times U \to X$. Then $E\underline{\times_X}F$ is naturally a fiber bundle over $X \times U$.

The set of k-jets of systems is the fiber product $J^kS = J^kF \times_{X\times U} J^kH$. Also, $J^kS^0 = J^kF\underline{\times_X}J^kH^0$. The sets J^kS^0 and $J^kH^0 \times U$ are also subbundles of J^kS and J^kH, respectively.

The evaluation jet mapping on the jet spaces is denoted by ev_k,

$$ev_k : S^r \times X \times U \to J^kS, \ (resp.\ S^{0,r} \times X \times U \to J^kS^0),$$

$$ev_k\big(\Sigma, x, u^{(0)}\big) = j^k\Sigma\big(x, u^{(0)}\big) = \big(j^kf\big(x, u^{(0)}\big), j^kh\big(x, u^{(0)}\big)\big),$$

where $j^k\Sigma(x, u^{(0)})$ denotes the k-jet of Σ at $(x, u^{(0)})$.

As we said, the statements of our results are the same for S^r and $S^{0,r}$. Our proofs also are the same because the dependence of h in the control plays no role in them.

In the following, we shall have to consider subspaces of S^r, F^r, H^r, or $S^{0,r}$, F^r, $H^{0,r}$, which will be denoted by the letters S, F, H, possibly with some additional indices, and which will have the following two properties:

- (A_1) S, F, H are subspaces of S^r, F^r, H^r, or $S^{0,r}$, F^r, $H^{0,r}$. They are Banach spaces for a stronger norm than the one from the overspaces S^r, F^r, ...
- (A_2) For all $(x_1, u_1) \neq (x_2, u_2)$, $(x_i, u_i) \in X \times U$, $i = 1, 2$, there is a $\Sigma \in S$ (resp. S^0) such that the k-jets $j^k\Sigma(x_i, u_i)$ at the points (x_i, u_i) have arbitrary prescribed values.

For the proof of our observability Theorems 2.1, 2.2, and 2.3 below, we will need only the obvious case where $S = S^r$ (resp. $S^{0,r}$). For the proof of our Theorem 2.4 (the real analytic case), we will make the choice $S = S_K$ or $S = S_K^0$, which will be described now (for the details, refer to [21]).

Here, $\tilde{X}$ denotes a complex manifold. O_X (resp. $O_{\tilde{X}}$) denotes the sheaf of germs of real analytic (resp. holomorphic) functions on X (resp. $\tilde{X}$). A complexification of (X, O_X) is a complex manifold $(\tilde{X}, O_{\tilde{X}})$ together with an antiinvolution (σ, σ^*) and a homomorphism $(\rho, \rho^*) : (X, O_X \otimes_R \mathbb{C}) \to (F, O_{\tilde{X}}|F)$, where F is the set of fixed points of σ.

Complexifications of real analytic Hausdorff paracompact manifolds do exist. Their germs along X are isomorphic. The complexification is a natural correspondence in the sense that $\tilde{X} \times \tilde{U}$ is a complexification of $X \times U$ in a natural way, and so is the tangent bundle $T\tilde{X}$ with respect to the tangent bundle TX of X.

We consider a complexification $\tilde{X}$ of X. By Grauert's theorem (see [21, 22]), $\tilde{X}$ contains a neighborhood of X, which is a **Stein** manifold. Replacing $\tilde{X}$ by this neighborhood, we can assume that $\tilde{X}$ is Stein. Then, $\mathbb{C}^{d_u}$ being the natural complexification of R^{d_u}, $\tilde{X} \times \mathbb{C}^{d_u}$ is also a Stein manifold, that we denote by $\widetilde{X \times U}$.

Let $\mathcal{K}$ be the class of closures of open, connected, σ-invariant, relatively compact neighborhoods of $X \times U$ in $\widetilde{X \times U}$. For $K \in \mathcal{K}$, let us denote by $BF(K)$ (resp. $VF(K)$), the set of all functions $h(x, u^{(0)})$ (resp. parametrized vector fields $f(x, u^{(0)})$) that are defined and continuous on K and holomorphic on $Int(K)$. The set $BF(K)$ endowed with the sup-norm becomes a Banach space. $VF(K)$ has also a Banach space structure, obtained by embedding $\tilde{X}$ into a complex space C^p, the embedding being compatible with the conjugation involutions on $\tilde{X}$ and C^p. The set $\widetilde{BF}(K)$ (resp. $\widetilde{VF}(K)$) is the real Banach subspace of conjugate-invariant elements of $BF(K)$ (resp. $VF(K)$).

The restriction $res_K : h \to h_{|X \times U}$ (resp. $f \to f_{|X \times U}$) maps $\widetilde{BF}(K)$ (resp. $\widetilde{VF}(K)$) continuously and injectively into $C^\infty(X \times U)$ (resp. $VF^\infty(X \times U)$). The subset $C^\omega(X \times U)$ (resp. $VF^\omega(X \times U)$) of all analytic functions (resp. parametrized vector fields) on $X \times U$ is the inductive limit over $\mathcal{K}$ ordered by inclusion, $\underrightarrow{Lim}(\widetilde{BF}(K))$ (resp. $\underrightarrow{Lim}(\widetilde{VF}(K))$).

This means that for any $h \in C^\omega(X \times U)$ (resp. $f \in VF^\omega(X \times U)$), $\exists K \in \mathcal{K}$, and $h^\circ \in \widetilde{BF}(K)$ (resp. $f^\circ \in \widetilde{VF}(K)$) such that $h = res_K(h^\circ)$ (resp. $f = res_K(f^\circ)$).

We denote by $S_K, F_K, H_K, S_K = F_K \times H_K, S_K \subset S^r$, the Banach spaces formed by the restrictions $res_K(\Sigma) = (res_K(f), res_K(h))$ of elements $\Sigma = (f, h)$ of $\widetilde{VF}(K) \times \widetilde{BF}(K)$, with the norm induced by $\widetilde{VF}(K) \times \widetilde{BF}(K)$.

Let us state some properties of S_K needed later on:

(i) the norm on S_K is stronger than the norm on S^r,

(ii) for any finite set of points $p_i = (x_i, u_i), 1 \le i \le N$ in $X \times U$, and elements $j_i \in J^k S(x_i, u_i)$, any $K \in \mathcal{K}$, there is a $\Sigma^\circ \in S_K$ such that $j^k \Sigma^\circ(x_i, u_i) = j_i$,

(iii) the evaluation map $ev_K : S_K \times X \times U \to J^k S$,

$$\left(\Sigma, x, u^{(0)}\right) \to j^k \Sigma\left(x, u^{(0)}\right),$$

is a C^∞ mapping.

Hence, in particular, the assumptions (A_1), (A_2) above are met. The spaces S_K^0, H_K^0, are also defined in the same way and have the same properties.

Let us end this section with a few more notations.

At a point x such that $f(x) = 0$, $T_X f(x)$ denotes the linearisation of f at x, i.e. the endomorphism of $T_x X$ associated to the vector field f

Denoting by ΔX the diagonal of $X \times X$, define the maps $DS\Phi_N$, $DS\Phi_N^\Sigma$, $D\Phi_N$, and $D\Phi_N^\Sigma$ (remember the notation $\underline{u}_N = (u^{(0)}, u^{(1)}, \ldots, u^{(N-1)})$):

$$
\begin{aligned}
&DS\Phi_N : ((X \times X)\backslash\Delta X) \times U \times R^{(N-1)d_u} \times S \to R^{Nd_y} \times R^{Nd_y} \times R^{Nd_u},\\
&DS\Phi_N(x_1, x_2, \underline{u}_N, \Sigma) = \left(\Phi_N^\Sigma(x_1, \underline{u}_N), \Phi_N^\Sigma(x_2, \underline{u}_N), \underline{u}_N\right),\\
&D\Phi_N : ((X \times X)\backslash\Delta X) \times U \times R^{(N-1)d_u} \times S \to R^{Nd_y} \times R^{Nd_y},\\
&D\Phi_N(x_1, x_2, \underline{u}_N, \Sigma) = \left(\Phi_N^\Sigma(x_1, \underline{u}_N), \Phi_N^\Sigma(x_2, \underline{u}_N)\right),
\end{aligned}
$$

$$
\begin{aligned}
&D\Phi_N^\Sigma(x_1, x_2, \underline{u}_N) = D\Phi_N(x_1, x_2, \underline{u}_N, \Sigma),\\
&DS\Phi_N^\Sigma(x_1, x_2, \underline{u}_N) = DS\Phi_N(x_1, x_2, \underline{u}_N, \Sigma).
\end{aligned}
$$

As we said, in the remainder of the chapter we will be interested in the following properties:

(F): $S\Phi_N^\Sigma$ is an embedding from $X \times U \times R^{(N-1)d_u}$ into $R^{Nd_y} \times R^{Nd_u}$.

Since X and U are compact, by the definition of $S\Phi_N^\Sigma$, this is equivalent to (F_1) and (F_2):

(F_1): $S\Phi_N^\Sigma$ is one-to-one,

(F_2): $S\Phi_N^\Sigma$ is an immersion.

(F_1) is equivalent to the fact that the map $D\Phi_N^\Sigma : ((X \times X)\backslash\Delta X) \times U \times R^{(N-1)d_u} \to R^{Nd_y} \times R^{Nd_y}$, avoids the diagonal in $R^{Nd_y} \times R^{Nd_y}$.

2. Statement of Our Differential Observability Results

In the next section, we shall prove the following theorems, that hold for $r > 0$, **large enough**.

Theorem 2.1. *The set of systems such that (F_2) is true, i.e., $S\Phi_N^\Sigma$ is an immersion,* ***contains an open dense*** *subset of S^r (resp. $S^{0,r}$), for $N \geq 2n$.*

Theorem 2.2. *The set of systems such that (F) is true (i.e., $S\Phi_N^\Sigma$ is an embedding, equivalently Σ is strongly differentially observable) contains a* ***residual*** *subset of S^r (resp. $S^{0,r}$), for $N \geq 2n+1$.*

A bound $B > 0$ on the derivatives of the controls being given, denote by I_B the interval $[-B, B]$.

Theorem 2.3. *The set of systems such that the restriction of $S\Phi_N^\Sigma$ to $X \times U \times I_B^{(k-1)d_u}$ is an embedding, is* ***open, dense*** *in S^r (resp. $S^{0,r}$), for $N \geq 2n+1$.*

Theorem 2.4. *(X analytic). The set of* **analytic** *systems such that $S\Phi_N^\Sigma$ is an embedding, is* **dense** *in S^r (resp. $S^{0,r}$), for $N \geq 2n+1$.*

Now, we shall give several examples showing that all these theorems are false when $d_y = d_u = 1$. In all the examples, $X = S^1$, the circle, and $U = [-1, 1]$.

Consider

$$(\Sigma^1)\begin{cases} \dot{\theta} = 1, \\ y = \varphi_1(\theta) + \varphi_2(\theta)u, \end{cases}$$

with the assumption (H): $\varphi_1'(\theta_0) = 0$, $\varphi_1''(\theta_0) \neq 0$, $\varphi_2'(\theta_0) \neq 0$.

One should observe that the condition (H) is stable under small perturbations and holds if $\theta_0 = 0$, $\varphi_1(\theta) = \cos(\theta)$, $\varphi_2(\theta) = \sin(\theta)$.

Exercise 2.1. Show that, if $\theta = \theta_0$, taking $u^{(0)} = 0$, we can compute $u^{(1)}, \ldots, u^{(N)}$ satisfying the equation

$$d_\theta \begin{pmatrix} y \\ \dot{y} \\ \cdot \\ \cdot \\ y^{(N)} \end{pmatrix} = 0,$$

and show that there exists an open neighborhood $\mathcal{U}$ of Σ^1 (C^2 open in S^∞), such that $S\Phi_k^\Sigma$ is not an immersion for any k, for $\Sigma \in \mathcal{U}$.

This is a counterexample of Theorems 2.1 and 2.2 when $d_y \leq d_u$. It is also a counterexample of Theorem 2.4.

A better example is the following system Σ_ε^1, for ε small:

$$(\Sigma_\varepsilon^1)\begin{cases} \dot{\theta} = 1, \\ y = \varepsilon\varphi_1(\theta) + \varphi_2(\theta)u, \end{cases}$$

with the same assumption (H) on θ_0, φ_1 and φ_2. Chose an arbitrary integer $k > 0$ and a real $B > 0$.

Exercise 2.2. Show that there is an ε_0 sufficiently small so that, for the system $\Sigma_{\varepsilon_0}^1$ and **for a C^k neighborhood $\mathcal{V}$ of** $\Sigma_{\varepsilon_0}^1$ in S^∞, there is a θ_0 and a point $u^{(0)}, u^{(1)}, \ldots, u^{(k-1)}$ such that $S\Phi_k^\Sigma$ is not immersive at $\theta_0, u^{(0)}, u^{(1)}, \ldots, u^{(k-1)}$, $u^{(0)} \in U$, $u^{(i)} \in I_B$, $1 \leq i \leq k-1$, $\Sigma \in \mathcal{V}$.

This is a counterexample to Theorem 2.3.

Using the same typology, we can also construct an example showing that, if $d_y > d_u$, the **set of systems Σ such that $S\Phi_N^{\Sigma}$ is an immersion is not open,** for any N. Consider the system:

$$(\Sigma_\varepsilon^2)\begin{cases}\dot{\theta} = 1,\\ y_1 = \varphi_1(\theta) + \varepsilon\varphi_2(\theta)u,\\ y_2 = 0,\end{cases}$$

with $\varphi_1(\theta) = \cos(\theta)$, $\varphi_2(\theta) = \sin(\theta)$. At $\theta_0 = 0$, the assumption (H) is satisfied.

Exercise 2.3. Show that, for $\varepsilon = 0$, $S\Phi_2^{\Sigma_\varepsilon^2}$ is an immersion. For $\varepsilon \neq 0$, ε small, $S\Phi_k^{\Sigma_\varepsilon^2}$ is not an immersion for any k, using the same reasoning as in the previous examples.

Exercise 2.4. Show that the mapping:

$$S^r \to C^{r-N+1}\left(X \times U \times R^{(N-1)d_u}, R^{Nd_y} \times U \times R^{(N-1)d_u}\right),$$
$$\Sigma \to S\Phi_N^{\Sigma},$$

is not continuous for the Whitney topology (over $C^{r-k+1}(X \times U \times R^{(N-1)d_u}, R^{kd_y} \times U \times R^{(N-1)d_u}))$.

Remark 2.1. Theorem 2.4 is not a consequence of Theorem 2.2: this theorem does not prove that there is an open dense subset of systems satisfying (F).

3. Proof of the Observability Theorems

The considerations of this section apply to both S^r and $S^{0,r}$. Moreover, in order to prove Theorem 2.4, relative to analytic systems, case which is crucial for the proof of Theorem 5.1 below, we will have to apply our reasonings to the cases where $S = S_K$ or $S = S_K^0$, for some $K \in \mathcal{K}$. To avoid cluttering our text with "respectively" and other alternatives, and for the sake of clarity, we will denote by S any of these classes of systems in all the proofs below, and H^0 will be considered as the subspace of H of all functions in H, independent of $u^{(0)}$. Below, when the difference between H and H^0 is not explicitly stated, $h(p)$ or $d_X^r h(p; \ldots)$ will mean $h(x_0)$ or $d_X^r h(x_0; \ldots)$, if $h \in H^0$ and $p = (x_0, u^{(0)})$, and all the expressions $d_X^p d_U^q h$ with $q > 0$ will be taken equal to 0. Note that the "bad sets" for $J^k S^0$ (see below) are just the intersections of the bad sets for $J^k S$ with $J^k S^0$.

3.1. Openness and Density of Immersivity

3.1.1. The "Bad Sets"

All our bad sets will be partially algebraic or semi-algebraic subbundles of vector bundles:

Definition 3.1. A subbundle B of a vector bundle E is said **partially algebraic (semi-algebraic)** (PA or PSA respectively), if its typical fiber is an algebraic (or semi-algebraic) subset of the vector space which is the typical fiber of E. It is equivalent to say that the fibers of the bundle are algebraic (resp. semi algebraic) in the corresponding fibers of E.

Definition 3.2.

(i) $\hat{B}_1(k)$ is the subset of $J^kS \times R^{(k-1)d_u}$ of all $(j^k\Sigma(x, u^{(0)}), u')$ such that: 1) $f(x, u^{(0)}) \neq 0$, 2) $rank(T_X\Phi_k^\Sigma(x, u^{(0)}, u')) < n$,

(ii) $\hat{B}_2(k)$ is the subset of $J^kS \times R^{(k-1)d_u}$ of all $(j^k\Sigma(x, u^{(0)}), u')$ such that: 1) resp. $f(x, u^{(0)}) = 0$, 2) the linear observed system $(d_Xh(x, u^{(0)}), T_X f(x, u^{(0)}))$ is observable, 3) $rank(T_X\Phi_k^\Sigma(x, u^{(0)}, u'))$ $< n$,

(**here**, $u' = u^{(1)}, \ldots, u^{(k-1)}$).

See Appendix 7.1 for the notion of observability for linear systems.

Definition 3.3. $B_3(k)$ is the subset of J^kS of all $j^k\Sigma(x, u^{(0)})$ such that:

(1) $f(x, u^{(0)}) = 0$ and (2) the linear observed system $(d_Xh(x, u^{(0)}),$ $T_X f(x, u^{(0)}))$ is unobservable.

Definition 3.4. $B_{im}(k)$ is the union $B_1(k) \cup B_2(k) \cup B_3(k)$, where $B_i(k)$, $i = 1, 2$ denotes the canonical projection of $\hat{B}_i(k)$ in J^kS.

The following two lemmas are obvious:

Lemma 3.1. *$\hat{B}_1(k)$, $\hat{B}_2(k)$, $B_1(k)$, $B_2(k)$, $B_3(k)$, $B_{im}(k)$ are PSA in their respective ambient vector bundles.*

Lemma 3.2. *$S\Phi_k^\Sigma$ is an immersion if and only if $j^k\Sigma$ avoids $B_{im}(k)$.*

3.1.2. Estimate of the Codimension of $B_{im}(k)$

Clearly, if B is a PSA subbundle of a vector bundle E, Codim(B, E), the codimension of B in E, is equal to the codimension of the typical fiber of B in the typical fiber of E.

a) Estimate of Codim$(B_3(k), J^k S)$: let $J^k S(p)$, $p = (x, u^{(0)})$, be a typical fiber of $J^k S$ and let $V \subset J^k S(p)$ be the vector subspace $V = \{j^k\Sigma(p) | f(p) = 0\}$. V has codimension n in $J^k S(p)$. Let $\lambda : V \to (T_x^* X)^{d_y} \times End(T_x X)$ be the mapping $\lambda(j^k\Sigma(p)) = (d_X h(p), T_X f(p))$. If $\mathbf{U} \subset (T_x^* X)^{d_y} \times End(T_x X)$ denotes the set of unobservable couples (C, A), then the typical fiber of $B_3(k)$ is $\lambda^{-1}(\mathbf{U})$. But, by Appendix 7.1, Codim$(\mathbf{U}, (T_x^* X)^{d_y} \times End(T_x X)) = d_y$.
Hence,

$$Codim(B_3(k; p), V) = d_y,$$

and

$$\begin{aligned} Codim(B_3(k; p), J^k S(p)) &= Codim(B_3(k; p), V) + Codim(V, J^k S(p)) \\ &= n + d_y. \end{aligned}$$

b) Estimate of Codim$(B_1(k), J^k S)$: let G be the open subset of $J^k F \underline{\times_X} TX \times R^{(k-1)d_u}$ of all $(j^k f(p), \Lambda, u')$ such that:

(i) $f(p) \neq 0$ (again, $p = (x, u^{(0)})$),
(ii) $\Lambda \neq 0$.

Let $\mu : J^k H \times_{X \times U} G$ (resp. $J^k H^0 \times_{X \times U} G) \to R^{kd_y}$, $\mu(j^k h(p), j^k f(p), \Lambda, u') = T_X \Phi_k^\Sigma(p, u')\Lambda$, where $\Sigma = (f, h)$.

The set G is the total space of an open subbundle of the vector bundle

$$J^k F \underline{\times_X} TX \times R^{(k-1)d_u} \to J^k F \times R^{(k-1)d_u},$$

with fibers the tangent spaces to X. The bundle G is conical in the sense that each fiber of G is a cone in the corresponding tangent space to X, and μ is a homogeneous submersion (see Appendix 7.2, Lemma 7.2). Hence, because $\hat{B}_1(k)$ is the canonical projection of $\mu^{-1}(0)$ in $J^k S \times R^{(k-1)d_u}$:

$$\begin{aligned} Codim\left(\mu^{-1}(0), J^k S \underline{\times_X} TX \times R^{(k-1)d_u}\right) &= kd_y, \\ Codim\left(\hat{B}_1(k), J^k S \times R^{(k-1)d_u}\right) &= kd_y + 1 - n, \\ Codim(B_1(k), J^k S) &= k(d_y - d_u) + 1 + d_u - n. \end{aligned}$$

c) Estimation of Codim$(B_2(k), J^k S)$ **(the most difficult case)**. $\hat{B}_2(k)$ is the subset of $J^k S \times R^{(k-1)d_u}$ of all $j^k\Sigma(p, u')$ such that: (1) $f(p) = 0$; (2) the linear system $(d_X h(p), T_X f(p))$ is observable, $p = (x_0, u^{(0)})$; (3) the tangent mapping $T_X \Phi_k^\Sigma(p, u') : T_{x_0} X \to R^{kd_y}$ has a nonzero kernel.

In the study of $\hat{B}_2(k)$, we will need the following lemma.

Lemma 3.3. *Let W be a finite dimensional vector space, $A : W \to W$, and $C : W \to R^l$ be two linear mappings. If there exists a nonzero vector $\Lambda \in W$ such that (i) $CA^r\Lambda = 0$ for $0 \leq r \leq t$, t some integer and (ii) the vectors $\{A^r\Lambda | 0 \leq r \leq t+1\}$ in W are linearly dependent, then the system (C, A) is not observable. (See Appendix 7.1 for the notion of observability of linear systems.)*

Proof. (ii) implies that there exists an integer m, $0 \leq m \leq t$, such that $A^{m+1}\Lambda$ belongs to the linear span, $Span\{A^i\Lambda | 0 \leq i \leq m\}$. Then, $A^r\Lambda \in Span\{A^i\Lambda | 0 \leq i \leq m\}$ for all integers r. Because $Span\{A^i\Lambda | 0 \leq i \leq m\} \subset Ker\, C$, $A^r\Lambda \in Ker\, C$ for all r. Hence, (C, A) is not observable. ■

To study $\hat{B}_2(k)$ more conveniently, we shall split it into several subsets that are easier to handle.

Definition 3.5. For every integer ρ, $1 \leq \rho \leq k-1$:

(i) $\hat{B}_6(k, \rho)$ will be the subsets of $\hat{B}_2(k)$ of all $(j^k\Sigma(p), u')$ such that: (1) $u^{(1)} = u^{(2)} = \ldots = u^{(\rho-1)} = 0$, (2) $d_U f(p, u^{(\rho)}) \neq 0$.
(ii) $\hat{B}_7(k, \rho)$ will be the subsets of $\hat{B}_2(k)$ of all $(j^k\Sigma(p), u')$ such that: (1) $u^{(1)} = u^{(2)} = \ldots = u^{(\rho-1)} = 0$, (2) $d_U f(p, u^{(\rho)}) = 0$.
(iii) Let $B_6(k, \rho)$, $B_7(k, \rho)$ denote the canonical projections of $\hat{B}_6(k, \rho)$, $\hat{B}_7(k, \rho)$, respectively, in J^kS.

Lemma 3.4.
(i) $\hat{B}_2(k) = \cup_{\rho=1}^{k-1}(\hat{B}_6(k, \rho) \cup \hat{B}_7(k, \rho))$,
$B_2(k, \rho) = \cup_{\rho=1}^{k-1}(B_6(k, \rho) \cup B_7(k, \rho))$,
(ii) $\hat{B}_6(k, \rho)$, $\hat{B}_7(k, \rho)$ *(resp.* $B_6(k, \rho)$, $B_7(k, \rho)$*) are PSA subbundles of* $J^kS \times R^{(k-1)d_u}$ *(resp.* J^kS*),*
(iii) $\hat{B}_6(k, \rho) = \hat{B}_7(k, \rho) = B_6(k, \rho) = B_7(k, \rho) = \varnothing$ *if* $\rho > n$.

Proof. (i) and (ii) are trivial. (iii): By Lemma 7.3, Appendix 7.2, for any $(j^k\Sigma(p), u') \in \hat{B}_6(k, \rho) \cup \hat{B}_7(k, \rho)$, there exists a $\Lambda \in T_{x_0}X$, $\Lambda \neq 0$, such that $d_X h(p; T_X f(p)^i\Lambda) = 0$ for $0 \leq i \leq \rho - 1$, $(p = (x_0, u^{(0)}))$. Applying Lemma 3.3 with $t = \rho - 1$, we get that $(d_X h(p), T_X f(p))$ is not observable because the vectors $T_X f(p)^i\Lambda$, $0 \leq i \leq \rho - 1$ are necessarily linearly dependent because $\dim T_{x_0}X = n < \rho$. This is a contradiction. ■

Now, we shall estimate the codimensions of $B_6(k, \rho)$ and $B_7(k, \rho)$.

d) Estimation of $\mathrm{Codim}(B_7(k, \rho), J^kS)$. Let $\tilde{B}_8(k, \rho)$ be the subset of $J^kS \underline{\times_X} P(TX) \times R^{d_u}$ of all $(j^k\Sigma(p), l, w)$ such that:

1. $f(p) = 0$,
2. $d_U f(p; w) = 0$,
3. $d_X h(p; T_X f(p)^i \Lambda) = 0$, $0 \leq i \leq \rho - 1$ and $d_X h(p; T_X f(p)^\rho \Lambda) + d_X d_U h(p; \Lambda \otimes w) = 0$, for all $\Lambda \in l$,
4. The vectors $T_X f(p)^i \Lambda$, $0 \leq i \leq \rho$ are linearly independent.

We claim that the canonical projection $B_8(k, \rho)$ of $\tilde{B}_8(k, \rho)$ in J^kS contains $B_7(k, \rho)$. An element $e \in B_7(k, \rho)$ is the image of an element $(j^k\Sigma(p), u') \in \hat{B}_7(k, \rho)$. By condition 3 in the definition of $\hat{B}_2(k)$, there exists a $\Lambda \in T_{x_0}X$, $\Lambda \neq 0$, such that $T_X \Phi_k^\Sigma(p, u')\Lambda = 0$.

$P(T_x X)$ is the projective space associated to $T_x X$. Let l be the class of Λ in $P(T_{x_0}X)$. If we show that the triple $(j^k\Sigma(p), l, u^{(\rho)})$ belongs to $\tilde{B}_8(k, \rho)$, it will follow that $B_8(k, \rho) \supset B_7(k, \rho)$.

Therefore, we have to check that the triple $(j^k\Sigma(p), l, u')$ satisfies the conditions 1-4 defining $\tilde{B}_8(k, \rho)$. But conditions 1 and 2 of the definition of $\tilde{B}_8(k, \rho)$ follow from condition 1 of the definition of $\hat{B}_2(k)$ and (ii)-2 of Definition 3.5 of $\hat{B}_7(k, \rho)$. Condition 3 follows from the fact that $T_X \Phi_k^\Sigma(p, u')\Lambda = 0$ and Lemma 7.3-(2), Appendix 7.2. Finally, condition 4 follows from the condition 3 just proven and Lemma 3.3 applied with $t = \rho - 1$, $W = T_{x_0}X$, $A = T_X f(p)$, $C = d_X h(p)$ and Λ. We get a contradiction with the observability condition 2 of Definition 3.2.

Clearly, $\tilde{B}_8(k, \rho)$, $B_8(k, \rho)$ are PSA subbundles of the bundles $J^kS \underline{\times_X} P(TX) \times R^{d_u}$ and J^kS respectively. The inclusion $B_8(k, \rho) \supset B_7(k, \rho)$ implies that

1. $\mathrm{Codim}(B_7(k, \rho), J^kS) \geq \mathrm{Codim}(B_8(k, \rho), J^kS)$.

   We have also:
2. $\mathrm{Codim}(B_8(k, \rho), J^kS) \geq \mathrm{Codim}(\tilde{B}_8(k, \rho), J^kS \underline{\times_X} P(TX) \times R^{d_u}) - d_u - n + 1$.

To estimate $\mathrm{Codim}(\tilde{B}_8(k, \rho), J^kS \underline{\times_X} P(TX) \times R^{d_u})$, let us note that $\tilde{B}_8(k, \rho)$ is the projection of the conical subset $\Psi^{-1}(0)$ of $J^kS \underline{\times_X} TX \times R^{d_u}$ where Ψ is a fiberwise mapping: $G \to pr_x^* TX \times R^{(1+\rho)d_y}$ defined as follows. The set G is the PSA subbundle of $J^kS \underline{\times_X} TX \times R^{d_u}$ of all $(j^k\Sigma(p), \Lambda, w)$ such that (i) $f(p) = 0$, and (ii) the vectors $T_X f(p)^i \Lambda$, $0 \leq i \leq \rho$, are linearly independent.

$$\Psi(j^k\Sigma(p), \Lambda, w) = (d_U f(p; w), d_X h(p; \Lambda), d_X h(p; T_X f(p)\Lambda), \ldots, d_X h(p; T_X f(p)^{\rho-1}\Lambda), d_X h(p; T_X f(p)^\rho \Lambda) + d_X d_U h(p; \Lambda \otimes w)).$$

Clearly, Ψ is a submersion because the $T_X f(p)^i \Lambda, 0 \leq i \leq \rho$ are linearly independent. Hence,

$$Codim(\Psi^{-1}(0), G) = n + (1+\rho)d_y.$$

Clearly, G is a submanifold of $J^k S\underline{\times_X} TX \times R^{d_u}$ of codimension n. Hence

$$Codim(\Psi^{-1}(0), J^k S\underline{\times_X} TX \times R^{d_u}) = 2n + (1+\rho)d_y,$$
$$Codim(\tilde{B}_8(k,\rho), J^k S\underline{\times_X} P(TX) \times R^{d_u}) = 2n + (1+\rho)d_y.$$

Using 1 and 2 above:

3. $Codim(B_7(k,\rho), J^k S) \geq n + (1+\rho)d_y - d_u + 1.$

e) Estimation of Codim$(B_6(k,\rho), J^k S)$. Let $G \subset J^k F\underline{\times_X} TX \times R^{(k-1)d_u}$ be the subset of all $(j^k f(p), \Lambda, u')$ such that: (i) $f(p) = 0$; (ii) $u^{(1)} = u^{(2)} = \ldots = u^{(\rho-1)} = 0$; and (iii) the vectors $T_X f(p)^i \Lambda, 0 \leq i \leq \rho$, are linearly independent; and (iv) $d_U f(p, u^{(\rho)}) \neq 0$.

The set G is clearly a PSA conical subbundle of $J^k F\underline{\times_X} TX \times R^{(k-1)d_u}$ and a submanifold of codimension:

4. $Codim\left(G, J^k F\underline{\times_X} TX \times R^{(k-1)d_u}\right) = n + (\rho-1)d_u.$

Let μ: $J^k H \times_{X\times U} G$ (resp. $J^k H^0 \times_{X\times U} G) \to R^{kd_y}$:

$$\mu(j^k h(p), j^k f(p), \Lambda, u') = T_X \Phi_k^{\Sigma}(p, u'; \Lambda), \text{ where } \Sigma = (f, h).$$

$\hat{B}_6(k,\rho)$ is contained in the image of $\mu^{-1}(0)$ by the canonical projection:

$$J^k S\underline{\times_X} TX \times R^{(k-1)d_u} \to J^k S \times R^{(k-1)d_u}.$$

To see this, let $(j^k\Sigma(p), u') \in \hat{B}_6(k,\rho)$. Then, by condition 3 of Definition 3.2, there is a $\Lambda \in T_{x_0}X$, $(p = (x_0, u^{(0)}))$, $\Lambda \neq 0$, such that $T_X \Phi_k^{\Sigma}(p, u'; \Lambda) = 0$. Let us show that $(j^k\Sigma(p), \Lambda, u')$ belongs to $\mu^{-1}(0)$. All we have to do is to check that $(j^k f(p), \Lambda, u') \in G$, where $\Sigma = (f, h)$. Now, $f(p) = 0$ by condition 1 of Definition 3.2 of $\hat{B}_2(k)$, $u^{(1)} = u^{(2)} = \ldots = u^{(\rho-1)} = 0$, and $d_U f(p, u^{(\rho)}) \neq 0$ by the conditions 1 and 2 of Definition 3.5 for $\hat{B}_6(k,\rho)$.

To check that the $T_X f(p)^i \Lambda$, $0 \leq i \leq \rho$ are linearly independent, note that by the Lemma 7.3 (2) of Appendix 7.2, the fact that $T_X \Phi_k^{\Sigma}(p; u', \Lambda) = 0$ implies that $d_X h(p; T_X f(p)^i \Lambda) = 0$ for $0 \leq i \leq \rho - 1$. Then, Lemma 3.3 and condition 2 of Definition 3.2 for $\hat{B}_2(k)$ imply that the $T_X f(p)^i \Lambda$,

$0 \le i \le \rho$, are linearly independent. Hence,

$$Codim\left(\hat{B}_6(k,\rho), J^k S \times R^{(k-1)d_u}\right) \ge$$
$$Codim\left(\mu^{-1}(0), J^k S \underline{\times_X} TX \times R^{(k-1)d_u}\right) + 1 - n.$$

Statement 1 of Lemma 7.3 in Appendix 7.2 implies immediately that μ is a submersion. Hence

$$Codim(\mu^{-1}(0), J^k H \times_{X \times U} G) = kd_y,$$
$$(\text{resp. } Codim(\mu^{-1}(0), J^k H^0 \times_{X \times U} G) = kd_y).$$

Using 4,

$$Codim\left(\mu^{-1}(0), J^k S \underline{\times_X} TX \times R^{(k-1)d_u}\right) = k_{d_y} + n + (\rho - 1)d_u,$$
$$Codim\left(\hat{B}_6(k,\rho), J^k S \times R^{(k-1)d_u}\right) \ge k_{d_y} + n + (\rho - 1)d_u - n + 1,$$
$$5.\ Codim(B_6(k,\rho), J^k S) \ge kd_y + 1 + (\rho - k)d_u.$$

Points 3, 5 and Lemma 3.4 (i) imply that

$$Codim(B_2(k), J^k S) \ge \min\{n + (1+\rho)d_y + 1 - d_u, kd_y + 1 + (\rho - k)d_u | \rho \ge 1\}.$$

Hence

$$Codim(B_2(k), J^k S) \ge \min\{n + 2d_y + 1 - d_u, k(d_y - d_u) + 1 + d_u\}.$$

f) Final estimation of $\mathrm{Codim}(B_{im}(k), J^k S)$. Combining the results of this section, a, b, c, d, and e, we obtain

$$Codim(B_{im}(k), J^k S) \ge \min\{n + d_y, k(d_y - d_u) + 1 - n + d_u, n + 2d_y + 1 - d_u, k(d_y - d_u) + 1 + d_u\}.$$

3.1.3. Proof of the Openness and Density, for Immersivity

Clearly, for $k \ge 2n$, $\mathrm{Codim}(B_{im}(k)) > n + d_u$, because $d_y > d_u$. The same applies to the closure $cl(B_{im}(k))$, which is still PSA (see [38]).

By Abraham's theorem (see [1]) on transversal density and openness of nonintersection, the set of $\Sigma \in S$ such that $j^k\Sigma$ avoids $cl(B_{im}(k))$ is open, dense. Of course, we have used our assumptions A_1, A_2 of Section 2, which imply that the map $j^k : S \to C^r(X \times U, J^k S)$ is a C^r representation, and $ev_k : S \times X \times U \to J^k S$ is a submersion.

Remark 3.1. In the case where $d_y = d_u$, $B_{im}(k)$ has codimension $d_u - n + 1$. This shows that the set on which $S\Phi_k^\Sigma$ is not immersive may have generically full codimension, whatever k. This is in agreement with the results of Chapter 3. (In that case, to get openness, we have to use an argument of openness of transversality to a closed stratified set, such as the one in [20].)

3.2. Density of Injectivity

3.2.1. The Bad Sets

Definition 3.6. Let $B_4(k)$ denote the subset of $J^kS \times J^kS$ of all couples $(j^k\Sigma(p), j^k\Sigma(q))$ such that: (1) $p \neq q$, $p = (x_1, u^{(0)})$, $q = (x_2, u^{(0)})$; (2) $f(p) = f(q) = 0$, and (3) $h(p) = h(q)$.

Definition 3.7.

(i) Let $\hat{B}_5(k)$ be the subset of $(J^kS)^2 \times R^{(k-1)d_u}$ of all tuples $((j^k\Sigma(p), j^k\Sigma(q), u')$ such that 1) $p \neq q$, $(p = (x_1, u^{(0)}),$ $q = (x_2, u^{(0)}))$, 2) $f(p) \neq 0$ or $f(q) \neq 0$, 3) $\Phi_k^\Sigma(p, u') = \Phi_k^\Sigma(q, u')$,

(ii) $B_5(k)$ denotes the canonical projection of $\hat{B}_5(k)$ in $(J^kS)^2$.

Again, we have the obvious lemmas:

Lemma 3.5. *$B_4(k)$, $\hat{B}_5(k)$, $B_5(k)$, are respectively PSA subbundles of $(J^kS)^2_*$, $(J^kS)^2_* \times R^{(k-1)d_u}$, $(J^kS)^2_*$, where $(J^kS)^2_*$ (resp. $(J^kS)^2_* \times R^{(k-1)d_u}$) denotes the restriction of $(J^kS)^2$ (resp. $(J^kS)^2 \times R^{(k-1)d_u}$) to $(X \times X \setminus \Delta X) \times \Delta U$, where ΔX and ΔU are the diagonals of $X \times X$ and $U \times U$ respectively.*

Lemma 3.6. *Let $Z \subset X \times X \setminus \Delta X$ and let $\Sigma \in S$ be such that the mapping; $Z \times U \to (J^kS)^2_*$, $(x, y, u^{(0)}) \to (j^k\Sigma(x, u^{(0)}), j^k\Sigma(y, u^{(0)}))$ avoids $B_{in}(k) = B_4(k) \cup B_5(k)$. Then $\Phi_k^\Sigma(x, u^{(0)}, u') \neq \Phi_k^\Sigma(y, u^{(0)}, u')$ for all $(x, y, u^{(0)}, u') \in Z \times U \times R^{(k-1)d_u}$.*

3.2.2. Estimation of the Codimension of $B_{in}(k)$[1]

a) Estimation of the codimension of $B_4(k)$ in $(J^kS)^2_*$: It is obvious that this codimension is $2n + d_y$.

b) Estimation of the codimension of $\hat{B}_5(k)$ in $(J^kS)^2_* \times R^{(k-1)d_u}$ and of $B_5(k)$ in $(J^kS)^2_*$: We will treat only the case $f(x, u^{(0)}) \neq 0$. The case $f(y, u^{(0)}) \neq 0$ is similar. Let $(x, y, u^{(0)}) \in (X \times X \setminus \Delta X) \times \Delta U$. The typical fiber $\hat{B}_5(k, x, y, u^{(0)})$ of $\hat{B}_5(k)$ in $J^kS(x, u^{(0)}) \times J^kS(y, u^{(0)}) \times R^{(k-1)d_u}$ is characterized by the following properties: (i) $f(x, u^{(0)}) \neq 0$ and (ii) $\Phi_k^\Sigma(x, u^{(0)}, u') - \Phi_k^\Sigma(y, u^{(0)}, u') = 0$.

[1] Do not confuse with $B_{im}(k)$ introduced in Definition 3.4.

Let G be the subset of $J^kS(x,u^{(0)}) \times J^kS(y,u^{(0)}) \times R^{(k-1)d_u}$ of all tuples $(j^k\Sigma(x,u^{(0)}), j^k\Sigma(y,u^{(0)}), u')$, such that $f(x,u^{(0)}) \neq 0$ and let $\chi : G \to R^{kd_y}$ be the mapping: $\chi(j^k\Sigma(x,u^{(0)}), j^k\Sigma(y,u^{(0)}), u') = \Phi_k^\Sigma(x,u^{(0)},u') - \Phi_k^\Sigma(y,u^{(0)},u')$. Then, $\hat{B}_5(k;x,y,u^{(0)},u') = \chi^{-1}(0)$. The map χ is an algebraic mapping, affine in $j^kh(x,u^{(0)})$.

By Lemma 7.1 in Appendix 7.2, for fixed $u' \in R^{(k-1)d_u}$ and $j^kf(x,u^{(0)})$, the linear mapping $j^kh(x,u^{(0)}) \in J^kH(x,u^{(0)}) \to \Phi^{j^k\Sigma}(x,u^{(0)},u')$ is surjective. This shows that the map $\chi : G \to R^{kd_y}$ is a submersion. Since $\hat{B}_5(k;x,y,u^{(0)}) = \chi^{-1}(0)$, $\mathrm{Codim}(\hat{B}_5(k;x,y,u^{(0)}), J^kS(x,u^{(0)}) \times J^kS(y,u^{(0)}) \times R^{(k-1)d_u}) = kd_y$.

Hence, $\mathrm{Codim}(\hat{B}_5(k), (J^kS)^2_* \times R^{(k-1)d_u}) = kd_y$. It follows that

$$Codim(B_5(k), (J^kS)^2_*) \geq k(d_y - d_u) + d_u.$$

c) Estimation of the codimension $\mathrm{Codim}(B_{in}(k), (J^kS)^2_*)$.

$$Codim(B_{in}(k), (J^kS)^2_*) \geq \min(2n + d_y, k(d_y - d_u) + d_u).$$

3.2.3. Proof of the Density of Injectivity

Let $k \geq 2n+1$. Then, $\mathrm{Codim}(B_{in}(k), (J^kS)^2_*) \geq 2n+1+d_u$. Therefore, another application of Abraham's transversal density theorem, (we omit the details), as in Section 3.1.3, shows that the set of all $\Sigma \in S$ such that $(j^k\Sigma(x,u^{(0)}), j^k\Sigma(y,u^{(0)}))$ avoids $Cl(B_{in}(k))$ for all $(x,y) \in Z \subset X \times X \setminus \Delta X$ and all $u^{(0)} \in U$ is residual. If Z is compact, it is open by openness of the nonintersection property on compact sets.

3.3. Proof of the Observability Theorems 2.1, 2.2, 2.3, and 2.4

Just applying the results of Sections 3.1.3 and 3.2, to the cases were $S = S^r$ (resp. $S = S^{0,r}$) proves Theorems 2.1 and 2.2.

Proof of Theorem 2.3. The set of embeddings of $(X \times U \times I_B^{(k-1)d_u})$ in $R^{kd_y} \times R^{kd_u}$ is open. On the other hand, the map

$$S^r \to C^{r-k+1}\left(X \times U \times I_B^{(k-1)d_u}, R^{kd_y} \times R^{kd_u}\right),$$
$$\Sigma \to \overline{S\Phi_k^\Sigma},$$

where $\overline{S\Phi_k^\Sigma}$ is the restriction of $S\Phi_k^\Sigma$ to $X \times U \times I_B^{(k-1)d_u}$, is continuous[2] (by compactness). ■

Hence, the set of $\Sigma \in S^r$ (resp. $\Sigma \in S^{0,r}$) such that $\overline{S\Phi_k^\Sigma}$ is an embedding is open. It is dense by Theorem 2.2.

[2] Compare with Exercise 2.4.

Proof of Theorem 2.4. We consider a fixed C^r system Σ^0. We approximate it in the C^r topology by an analytic one, Σ^1. By Section 1, Σ^1 is in one of our Banach spaces S_K (resp. S_K^0). The Banach space S_K satisfies the assumptions A_1 and A_2 of Section 1, by the properties (i), (ii), and (iii) of the same section. We just apply the results of Sections 3.1.3 and 3.2.3 to the case in which $S = S_K$ (resp. S_K^0), to get the result. ■

4. Equivalence between Observability and Observability for Smooth Inputs

In this section, we consider **analytic systems**, and we prove that, for these systems, C^ω – observability (i.e., observability for all C^ω inputs) implies observability (i.e., observability for all L^∞ inputs). In fact, these notions are equivalent. This result will be crucial in order to prove our final **approximation theorem** (Theorem 5.1) in Section 5.[3] To prove this, we need some technical tools, mainly, the tangent cone to a subanalytic set.

Exercise 4.1. Show that, in the C^∞ case, observability and C^ω observability are not equivalent properties.

4.1. Preliminaries

We start with X, a real analytic manifold; $Z \subset X$, a subanalytic subset; Z_{reg}, the subset formed by the regular points of Z (i.e., the points $x \in Z$ having a neighborhood O in X, such that $O \cap Z$ is a real analytic connected manifold). The set Z_{reg} is a disjoint union of analytic manifolds, and it is open and dense in Z. The subset Z_{reg} is also subanalytic in X (see Appendix, Section 1.3). This is a key point for the proof of the facts P_1 to P_6 below.

Let us now define $TC(Z)$, the tangent cone to Z: TZ_{reg}, the tangent bundle to Z_{reg} is well defined because Z_{reg} is smooth. Denote by $\Pi_X : TX \to X$ the canonical projection. We define

$$TC(Z) = \Pi_X^{-1}(Z) \cap \overline{TZ_{reg}},$$

where $\overline{TZ_{reg}}$ is the closure of TZ_{reg} in TX.

Now, let us state the main properties of $TC(Z)$:

P_1: $TC(Z)$ is subanalytic in TX. If Z is closed then $TC(Z) = \overline{TZ_{reg}}$,

P_2: If Z is smooth, $TC(Z) = TZ$,

P_3: If $Y \subset Z$ is subanalytic **open** in Z, then the restriction:

$$TC(Z)_{|Y} = TC(Y).$$

[3] There is a related result in the paper [46], based upon desingularization techniques.

$TC(Z)_{|Y} = \Pi_Z^{-1}(Y)$, where $\Pi_Z : TC(Z) \to Z$ denotes the canonical projection (restriction of $\Pi_X : TX \to X$ to $TC(Z)$). In particular, because Z_{reg} is open in Z,

$$TC(Z)_{|Z_{reg}} = TC(Z_{reg}) = TZ_{reg}.$$

P_4: If $Z = \{Z_\alpha | \alpha \in A\}$ is a stratification of Z satisfying the condition (a) of Whitney, (see Appendix, Section 1.4), then $TC(Z) \supset TZ_\alpha$ for all $\alpha \in A$.

P_5: If $x : \delta \to X$ is an absolutely continuous curve on the interval δ, the image of which is contained in Z, then the set of $t \in \delta$, such that $\frac{dx(t)}{dt}$ exists and is contained in $TC(Z)$, has full measure.

P_6: The dimension of $TC(Z)$ is $2\dim(Z)$.

1. Proof of P_1 (see also [13]). The question being local, we can assume that $X = R^n$, $TX = X \times X$, $Z \subset X$. $\Pi_X^{-1}(Z) = Z \times X$ is therefore subanalytic as a product of subanalytic sets. Hence, it is sufficient to show that $\overline{TZ_{reg}}$ is subanalytic. As a consequence, all we have to prove is that if Z is subanalytic and smooth, then TZ is subanalytic.

Let S be the unit sphere of $X = R^n$. Consider I_{nc}, the subset of $X \times X \times S$ defined by $I_{nc} = \{(x, y, v) | x, y \in X, v \in S, (y - x) \wedge v = 0, < (y - x), v > \geq 0\}$, (here, $< ., . >$ denotes the Euclidean scalar product on R^n). This subset I_{nc} is semi-analytic, it is also a smooth submanifold of $X \times X \times S$ with a boundary. Let p: $I_{nc} \to X \times X$ be the restriction to I_{nc} of the canonical projection $\mathrm{Pr}_{1,2}$: $X \times X \times S \to X \times X$. The set $Z \times Z$ is subanalytic in $X \times X$. The diagonal ΔX of $X \times X$ is analytic in $X \times X$. Therefore, $Z_*^{(2)} = Z \times Z \setminus (\Delta X \cap Z \times Z) = Z \times Z \setminus \Delta Z$ is subanalytic. Hence, $p^{-1}(Z_*^{(2)}) = \mathrm{Pr}_{1,2}^{-1}(Z_*^{(2)}) \cap I_{nc} = Sec(Z)$ is still subanalytic. Set $Dir(Z) = \overline{Sec(Z)} \cap p^{-1}(\Delta Z)$. Both $\overline{Sec(Z)}$ and $p^{-1}(\Delta Z) = \{(z, z, v) | z \in Z, v \in S\}$ are subanalytic, hence so is $Dir(Z)$. ($Dir(Z)$ is nothing but the set of (z, z, v) such that v is tangent to Z at z: $(z, z, v) \in Dir(Z)$ iff there is a sequence $(x_n, y_n) \to (z, z)$, $x_n \neq y_n$, $x_n, y_n \in Z$, $\frac{y_n - x_n}{\|y_n - x_n\|} \to v$).

Consider now $\mu : X \times X \times S \times R \to X \times X = TX$,

$$\mu(x, y, v, t) = (x, tv).$$

Restricted to $\Delta X \times S \times R$, μ is proper. $R^+ \subset R$ is semi-analytic, therefore, $\mu(Dir(Z) \times R^+)$ is subanalytic. But, clearly, $\mu(Dir(Z) \times R^+) = TC(Z)$. Hence, $TC(Z)$ is subanalytic.

2. Proof of P_2. If Z is smooth, $Z_{reg} = Z$, $TZ_{reg} = TZ$.

$$TC(Z) = \overline{TZ_{reg}} \cap \Pi_X^{-1}(Z) = \overline{TZ} \cap \Pi_X^{-1}(Z) = TZ.$$

3. Proof of P_3. $TC(Z)_{|Y} = \Pi_Z^{-1}(Y) = \Pi_X^{-1}(Y) \cap TC(Z) = \Pi_X^{-1}(Y) \cap \Pi_X^{-1}(Z) \cap \overline{TZ_{reg}} = \Pi_X^{-1}(Y) \cap \overline{TZ_{reg}}$. Otherwise, $TC(Y) = \Pi_X^{-1}(Y) \cap \overline{TY_{reg}}$.

But, $Y_{reg} = Z_{reg} \cap Y$, because Y is open in Z. Hence, $TY_{reg} \subset TZ_{reg}$, $\overline{TY_{reg}} \subset \overline{TZ_{reg}}$. Therefore, $TC(Y) \subset TC(Z)_{|Y}$.

Conversely, if $\xi_y \in \Pi_X^{-1}(Y) \cap TC(Z)$, $y \in Y$, then there is a sequence $\xi_{y_n} \to \xi_y, \xi_{y_n} \in TZ_{reg}, y_n \in Z_{reg}, y_n \to y$. For n sufficiently large, because Y is open in Z, $y_n \in Y_{reg}$, $\xi_{y_n} \in TY_{reg}$, hence, $\xi_y \in \overline{TY_{reg}}$. At the end, $\xi_y \in \Pi_X^{-1}(Y) \cap \overline{TY_{reg}} = TC(Y)$.

4. Proof of P_4. Let Z_α be a maximal stratum of Z (i.e., there is no Z_β, $Z_\alpha \neq Z_\beta$, $Z_\alpha \subset \overline{Z_\beta}$). Then, $Z_\alpha \subset Z_{reg}$ and by P_3, $TZ_\alpha \subset TZ_{reg} \subset TC(Z)$. If Z_α is not maximal, i.e., $Z_\alpha \subset \overline{Z_\beta}$, then, by "Whitney condition (a)," $T(Z_\alpha) \subset \overline{T(Z_\beta)}$. Hence, $T(Z_\alpha) \subset \overline{TZ_{reg}}$. Also, because Z_α is smooth, $T(Z_\alpha) \subset \Pi_X^{-1}(Z_\alpha)$. Finally, $T(Z_\alpha) \subset TC(Z)$.

5. Proof of P_5. Let $\{Z_\alpha | \alpha \in A\}$ be a Whitney stratification of Z by analytic manifolds that are also subanalytic in X. For each $\alpha \in A$, let δ_α be the set of $t \in \delta$ such that $x(t) \in Z_\alpha$. The set of points of δ_α that are not accumulation points of δ_α is countable and hence has measure zero. Let $\widehat{\delta_\alpha}$ be the set of $t \in \delta_\alpha$ that are accumulation points of δ_α. The set of t such that $\frac{dx(t)}{dt}$ exists has full measure in δ. Pick such a t in δ_α for some α. If $t \in \widehat{\delta_\alpha}$, then $\frac{dx(t)}{dt} \in T_{x(t)}Z_\alpha$ because Z_α is smooth. But we know by P_4 that $T(Z_\alpha) \subset TC(Z)$. Hence, $\frac{dx(t)}{dt} \in TC(Z)$.

6. Proof of P_6. (For the concept of dimension, see Appendix, Section 1.3). $\dim(Z) = \dim(Z_{reg})$. $\dim(TC(Z)) = \dim(TZ_{reg}) = 2\dim(Z_{reg})$. ■

4.2. Back to Observability

We consider an analytic system, just as in the previous sections:

$$(\Sigma)\ \dot{x} = f(x, u), \ \ y = h(x, u).$$

The considerations below are valid whether or not h depends on u.

We just assume that u takes values in a semi-analytic compact set U (which is the case for the systems considered previously, $U = I^{d_u}$, if I is a compact interval). The output space Y is just assumed to be an analytic manifold, **X is not assumed to be compact.**

Let $X_*^{(2)} = X \times X \setminus \Delta X$, ΔX being the diagonal of $X \times X$, and let $h \times h: X \times X \times U \to Y \times Y$ be the mapping $(x_1, x_2, u) \to (h(x_1, u), h(x_2, u))$. Now, we define two decreasing sequences of subanalytic subsets of $X_*^{(2)} \times U$ and $X_*^{(2)}$ respectively, as follows:

$\hat{\Gamma}_0 = (X_*^{(2)} \times U) \cap (h \times h)^{-1}(\Delta Y) = \{(x_1, x_2, u) | h(x_1, u) = h(x_2, u)\}$, $\Gamma_0 = \mathrm{Pr}_{1,2}(\hat{\Gamma}_0)$, $\mathrm{Pr}_{1,2} : X \times X \times U \to X \times X$ is the projection $(x_1, x_2, u) \to (x_1, x_2)$. $\mathrm{Pr}_{1,2}$ is a proper map. It is clear that $\hat{\Gamma}_0$ is semi-analytic in $X \times X \times U$, closed in $X_*^{(2)} \times U$. The set Γ_0 is subanalytic in $X \times X$, closed in $X_*^{(2)}$, which is itself an analytic manifold.

Assume that $\hat{\Gamma}_i$, Γ_i have been defined for $i \leq j$, $\hat{\Gamma}_i$ subanalytic closed in $X_*^{(2)} \times U$, Γ_i subanalytic closed in $X_*^{(2)}$.

Let us denote by $\hat{f} : X_*^{(2)} \times U \to TX_*^{(2)} \subset TX \times TX$, the analytic map: $(x_1, x_2, u) \to (f(x_1, u), f(x_2, u))$. Then, we define $\hat{\Gamma}_{j+1} = \hat{f}^{-1}(TC(\Gamma_j)) \cap \hat{\Gamma}_j$, $\Gamma_{j+1} = \mathrm{Pr}_{1,2}(\hat{\Gamma}_{j+1})$.

It is clear that $\hat{\Gamma}_{j+1}$ is subanalytic closed $X_*^{(2)} \times U$ and Γ_{j+1} is subanalytic closed in $X_*^{(2)}$. Also:

$$\hat{\Gamma}_0 \supset \hat{\Gamma}_1 \supset \ldots \supset \hat{\Gamma}_i \supset \ldots$$
$$\Gamma_0 \supset \Gamma_1 \supset \ldots \supset \Gamma_i \supset \ldots$$

We will show that as soon as $i \geq 2n + 1 = N$, $\Gamma_i = \Gamma_N$ and $\hat{\Gamma}_i = \hat{\Gamma}_{N+1}$ for $i \geq N + 1$.

First of all, let us recall the notion of **local dimension** [25]: If Z is a subanalytic subset of a submanifold X, and if $x \in X$, $\dim_x(Z) = Inf_V\{\dim(V \cap Z_{reg}) |\ V$ is an open neighborhood of $x\}$, with the convention that $\dim \varnothing = -1$. If $Z' \subset Z$ is a subanalytic subset of X, then for all $x \in X$:

$$\dim_x(Z') \leq \dim_x(Z).$$

We need the following fundamental lemma:

Lemma 4.1. *If for some integer $m \in \{-1, 0, 1, 2, \ldots\}$ and for some $z \in X_*^{(2)}$, $\dim_z(\Gamma_m) = \dim_z(\Gamma_{m+1})$, then for all $j \geq m$, $\dim_z(\Gamma_j) = \dim_z(\Gamma_m)$.*

The proof of this lemma is postponed to Section 4.3.

Corollary 4.2. *As soon as $i > \dim(\Gamma_0)$, $\Gamma_{i+1} = \Gamma_i$.*

Proof. By Lemma 4.1, for all $z \in X_*^{(2)}$, as soon as $i > \dim_z(\Gamma_0)$, $\dim_z(\Gamma_{i+1})= \dim_z(\Gamma_i)$. Because $\dim(\Gamma_0) = \sup_{z \in X_*^{(2)}} \dim_z(\Gamma_0)$, $\dim_z(\Gamma_{i+1}) = \dim_z(\Gamma_i)$ for all $z \in X_*^{(2)}$ and all $i > \dim(\Gamma_0)$. This implies that $\Gamma_{i+1} = \Gamma_i$, because the Γ_i are closed in $X_*^{(2)}$. ■

Corollary 4.3. *As soon as $i > 2n + 1$, $\hat{\Gamma}_{i+1} = \hat{\Gamma}_i$.*

Proof. $\hat{\Gamma}_{i+1} = \hat{f}^{-1}(TC(\Gamma_i)) \cap \hat{\Gamma}_i$. Because $i > \dim(\Gamma_0) + 1$, Corollary 4.2 shows that $\Gamma_i = \Gamma_{i-1}$. Therefore, $\hat{\Gamma}_{i+1} = \hat{f}^{-1}(TC(\Gamma_{i-1})) \cap \hat{\Gamma}_i = \hat{f}^{-1}(TC(\Gamma_{i-1})) \cap \hat{\Gamma}_{i-1}$, because $\hat{\Gamma}_i = \hat{f}^{-1}(TC(\Gamma_{i-1})) \cap \hat{\Gamma}_{i-1}$. Hence, $\hat{\Gamma}_{i+1} = \hat{\Gamma}_i$. ■

Setting $\Gamma = \cap_{i \geq 0} \Gamma_i$, and $\hat{\Gamma} = \cap_{i \geq 0} \hat{\Gamma}_i$, Corollary 4.3 implies obviously the following proposition:

Proposition 4.4. *(i)* $\Gamma = \Gamma_m$ *for* $m > 2n$ *and (ii)* $\hat{\Gamma} = \hat{\Gamma}_m$ *for* $m > 2n + 1$.

The basic fact about Γ is:

Proposition 4.5. *Assume that there exist two trajectories* $(\bar{x}_i, \bar{u}) : [0, \bar{T}] \to X \times U$, $i = 1, 2$, *of the system* Σ *such that* $\bar{x}_1(0) \neq \bar{x}_2(0)$ *and* $h(\bar{x}_1, \bar{u}) = h(\bar{x}_2, \bar{u})$ *for almost all* $t \in [0, \bar{T}]$. *Then* $\Gamma \neq \varnothing$ *and* $\hat{\Gamma} \neq \varnothing$.

The proof of Proposition 4.5 is also postponed to Section 4.3.

By definition and by Proposition 4.4, Γ and $\hat{\Gamma}$ satisfy:

(a) $\hat{f}(\hat{\Gamma}) \subset TC(\Gamma)$,
(b) $\Gamma = \mathrm{Pr}_{1,2}(\hat{\Gamma})$.

Pick any $(x_1, x_2) \in \Gamma_{reg}$. There is a u such that $(x_1, x_2, u) \in \hat{\Gamma}$. (a) and (b) above and the property P_3 of $TC(\Gamma)$ imply that, for any such $u \in U$:

(c) $\hat{f}(x_1, x_2, u) \in T_{(x_1,x_2)}\Gamma_{reg}$.

Also, because $\mathrm{Pr}_{1,2}(\hat{\Gamma}) = \Gamma$ and $\mathrm{Pr}_{1,2}$ is proper, it is known (see Appendix, Section 1.4) that the restriction $\mathrm{Pr}_{1,2|\hat{\Gamma}}$ can be stratified, that is: for any stratum S of Γ, $\hat{\Gamma} \cap \mathrm{Pr}_{1,2}^{-1}(S)$ is a union of strata in $\hat{\Gamma}$ and the restriction of $\mathrm{Pr}_{1,2}$ to such a stratum $\hat{S}$, $\mathrm{Pr}_{1,2|\hat{S}}$, is an analytic submersion. We pick two such strata S and $\hat{S}$, such that S has maximum dimension $\dim(\Gamma)$ (so that S is contained in Γ_{reg}). Considering the submersion $\mathrm{Pr}_{1,2|\hat{S}} : \hat{S} \to S$, we see that we can chose an analytic section

$$\hat{u} : V \to \hat{S}, \ \hat{u}(x_1, x_2) = (x_1, x_2, u(x_1, x_2)),$$

where V is an open subset of S.

Also, by (c) above, we know that, for all $(x_1, x_2) \in V$, $H(x_1, x_2) = (f(x_1, u(x_1, x_2)), f(x_2, u(x_1, x_2))) \in T_{(x_1,x_2)}V$. So H is an analytic vector field on V. Let $\gamma : [0, T] \to V$, $\gamma = (x_1(t), x_2(t))$, be an integral curve of

H in V. Setting $v(t) = u(x_1(t), x_2(t))$ for $t \in [0, t]$, one has by construction: $h(x_1(t), v(t)) = h(x_2(t), v(t))$ because $\hat{u}(V) \subset \hat{S} \subset \hat{\Gamma}_0 = \{(x_1, x_2, u) \mid h(x_1, u) = h(x_2, u)\}$. On the other hand, $x_1(t)$ and $x_2(t)$ are, by construction, distinct trajectories of our system Σ, relative to the input $v(t)$. Hence, $v(t)$ is an analytic function defined on the time interval $[0, T]$, which "makes our system Σ unobservable."

Finally, we have shown that, if our system is **unobservable** for some input, then it is also **unobservable** for another C^ω input. This conclusion is summarized in the next theorem.

Theorem 4.6. *For an **analytic** system Σ, (either $\Sigma \in S^\omega$ or $\Sigma \in S^{0,\omega}$), the following properties are equivalent:*

(i) Σ is observable for all L^∞ inputs.
(ii) Σ is observable for all C^ω inputs.

4.3. Proof of Lemma 4.1 and Proposition 4.5.

1. Proof of Lemma 4.1.

It is sufficient to show that $\dim_z \Gamma_{m+2} = \dim_z \Gamma_{m+1}$. There is an open neighborhood V_0 of z in X such that for every open neighborhood V of z contained in V_0, $\dim(V \cap \Gamma_{i,reg}) = \dim_z(\Gamma_i)$ for $i = m, m+1, m+2$. Because $\dim_z \Gamma_m = \dim_z \Gamma_{m+1}$, $V \cap \Gamma_{m+1}$ contains a submanifold ω_V of dimension $\dim_z \Gamma_m$, open in Γ_m. If $y \in \omega_V$, there is a $u_y \in U$ such that $(y, u_y) \in \hat{\Gamma}_{m+1}$, since $y \in \Gamma_{m+1}$. Therefore, $\hat{f}(y, u_y) \in TC_y(\Gamma_m)$. By the property P_2 Section 4.1, $TC_y(\Gamma_m) = TC_y(\omega_V) = T_y(\omega_V) = TC_y(\Gamma_{m+1})$. (Because $\omega_V \subset \Gamma_{m+1} \subset \Gamma_m$, ω_V is also open in Γ_{m+1}.) Finally

$$\hat{f}(y, u_y) \in TC(\Gamma_{m+1})$$
$$(y, u_y) \in \hat{\Gamma}_{m+1},$$

which shows that $(y, u_y) \in \hat{\Gamma}_{m+2}$ and hence, $y \in \Gamma_{m+2}$.

Therefore, for each open neighborhood V of z contained in V_0, $\omega_V \subset \Gamma_{m+2}$. On the other hand, there exists an open neighborhood W of z, $W \subset V_0$, such that $\dim_z(\Gamma_{m+2}) = \dim(W \cap \Gamma_{m+2})$. Because $\omega_V \subset \Gamma_{m+2}$ and $\dim(W \cap \Gamma_{m+1}) = \dim_z(\Gamma_{m+1})$ (by the definition of V_0), one has

$$\begin{aligned}\dim_z \Gamma_m &= \dim \omega_V \leq \dim_z \Gamma_{m+2} = \dim W \cap \Gamma_{m+2} \\ &\leq \dim W \cap \Gamma_{m+1} = \dim_z \Gamma_{m+1}.\end{aligned}$$

Finally, $\dim_z \Gamma_m = \dim_z \Gamma_{m+2} = \dim_z \Gamma_{m+1}$.

2. *Proof of Proposition 4.5*

Since the $\bar{x}_i$, $i = 1, 2$ are continuous and $\bar{x}_1(0) \neq \bar{x}_2(0)$, there exists a T, $0 < T < \bar{T}$ such that $\bar{x}_1(t) \neq \bar{x}_2(t)$ for all $t \in [0, T]$. Let us denote by $\bar{z}$: $[0, T] \to X_*^{(2)}$ the curve $\bar{z}(t) = (\bar{x}_1(t), \bar{x}_2(t))$. We will show by induction on i that for almost all $t \in [0, T]$, $(\bar{z}(t), \bar{u}(t)) \in \hat{\Gamma}_i$. This will imply the proposition.

For $i = 0$, it results from the assumption of the proposition. Assume that it is proven for $i \leq m$. Then, $\bar{z}$ is an absolutely continuous curve taking its values in Γ_m. By the property P_5, Section 4.1, for almost all $t \in [0, T]$:

$$\frac{d\bar{z}}{dt} \in TC(\Gamma_m),$$

this shows that for almost all $t \in [0, T]$,

$$\hat{f}(\bar{z}(t), \bar{u}(t)) \in TC(\Gamma_m),$$

hence, for almost all $t \in [0, T]$, $(\bar{z}(t), \bar{u}(t)) \in \hat{f}^{-1}(TC(\Gamma_m)) \cap \hat{\Gamma}_m = \hat{\Gamma}_{m+1}$.

■

5. The Approximation Theorem

Recall that, if Σ (analytic), is as in the previous Sections 2 and 3, such that $S\Phi_k^\Sigma$ is an embedding, then Σ is observable for all C^ω inputs (Σ is observable for all C^k inputs, which is stronger):

A C^k input $u(t)$ being given on some interval $[0, \Theta]$, and $x_1, x_2, x_1 \neq x_2$ being given initial conditions, assume that the corresponding outputs are equal on some time subinterval $[0, \tau]$. Then their $k - 1$ first derivatives at time zero are also equal, and they, with the $u^{(j)}(0)$, are just the components of $S\Phi_k^\Sigma$ by definition. This is impossible as $S\Phi_k^\Sigma$ is injective. The system Σ is observable for u.

In fact, the fact that $S\Phi_k^\Sigma$ is an embedding (strong differential observability) expresses that $P_{\Sigma,u}$, the *initial* $-$ *state* $\to$ *output* $-$ *trajectory* mapping, is an embedding for all of the considered k-times differentiable inputs u. Although, Σ observable only means that $P_{\Sigma,u}$ is injective, but for all L^∞ inputs u.

The results of Sections 2 and 4 show that for $d_y > d_u$:

1. Any C_r system Σ^0 can be approximated by an analytic one Σ^1, which is observable for all C^ω inputs (for all C^{2n+1} inputs): Theorem 2.4.
2. Σ^1 is in fact observable (for all L^∞ inputs): Theorem 4.6.

Therefore:

Theorem 5.1. ***(Approximation by observable analytic systems).*** *Any system $\Sigma^0 \in S^r$ (resp. $S^{0,r}$), r sufficiently large, can be approximated by an* *observable one $\Sigma^1 \in S^r$ (resp. $S^{0,r}$) (observable for all L^∞ inputs), that moreover can be chosen analytic and such that $S\Phi_k^{\Sigma^1}$ is an embedding, for some k.*

6. Complements

The two following results are important. The first one concerns uncontrolled systems, i.e., Σ is of the form

$$(\Sigma_{uc})\ \frac{dx}{dt} = f(x),\ y = h(x).$$

The manifold X is again assumed to be compact. Then, the following theorem holds.

Theorem 6.1. *The set of uncontrolled systems Σ_{uc} that are strongly differentially observable, of order $N = 2n + 1$, (i.e., Φ_N^Σ is an embedding) is open, dense.*

Exercise 6.1. Prove Theorem 6.1.

This is easy, as a consequence of the main theorems in this chapter. For a direct proof, see [16].

The second important result concerns the class of control affine systems. These systems are very common in practice. Recall that they are of the form

$$(\Sigma_{ca})\ \frac{dx}{dt} = f(x) + \sum_{i=1}^{d_u} g_i(x)u_i,\ y = h(x).$$

The following theorem holds.

Theorem 6.2. *The Theorems 2.1, 2.2, 2.3, and 2.4 are all true in the class of control affine systems.*

Exercise 6.2. Prove Theorem 6.2.

This exercise is not that easy, although the general idea of the proof is the same. This has been done in [3].

Also, the following interesting result holds: X is again an analytic compact manifold. Let us say that **a vector field f on X is observable** if there exists a continuous function $h : X \to R$, such that $\Sigma = (f, h)$ is observable.

Theorem 6.3.

*(1) An **analytic** vector field f is observable iff it has only **isolated singularities.***

(2) If (1) holds, then, the set of analytic maps h such that $\Sigma = (f, h)$ is observable, is dense in H^r.

Exercise 6.3. Prove Theorem 6.3.

This is not an easy exercise. For hints, see [31].

7. Appendix

7.1. Unobservable Linear Systems

Let K be a field. For our purposes, we need $K = R$ or $\mathbb{C}$ (real or complex numbers). Let L_K denote the set of all linear systems with observations, i.e., the vector space $Lin(K^n, K^{d_y}) \times End(K^n)$. For $L \in L_K$, $L = (C, A)$, denote by I_L the subspace $\bigcap_{i \geq 0} \ker CA^i$ of K^n.The subspace I_L is invariant under A. The system L is called *unobservable* if I_L is not reduced to $\{0\}$. Denote by U_K the subset of L_K formed by these systems. It is easy to see that there exists a universal family of polynomials $P_1, \ldots, P_r$, in variables representing the coefficients of the matrices C and A, with coefficients in $\mathbb{Z}$, such that U_K is an affine variety, i.e., the set of points of L_K that are the common zeros of this family. This implies that

$$U_R = U_{\mathbb{C}} \cap L_R,$$

and that

$$\text{Codim}_R(U_R, L_R) \geq \text{Codim}_{\mathbb{C}}(U_{\mathbb{C}}, L_{\mathbb{C}}). \tag{35}$$

We want to compute $\text{Codim}_R(U_R, L_R)$. To do this, let us introduce the space $V_K = \{(C, A)|\ C$ annihilates an eigenvector of A in $K^n\}$. Clearly, $V_K \subset U_K$. Hence, $\text{Codim}_K(V_K, L_K) \geq \text{Codim}_K(U_K, L_K)$. However, for $K = \mathbb{C}$, these codimensions are actually equal because $V_{\mathbb{C}} = U_{\mathbb{C}}$: Any vector subspace of $\mathbb{C}^n$ invariant by A and not reduced to $\{0\}$ contains an eigenvector of A. Hence we have, using (35):

$$\begin{aligned} \text{Codim}_R(V_R, L_R) &\geq \text{Codim}_R(U_R, L_R) \geq \text{Codim}_{\mathbb{C}}(U_{\mathbb{C}}, L_{\mathbb{C}}) \\ &= \text{Codim}_{\mathbb{C}}(V_{\mathbb{C}}, L_{\mathbb{C}}). \end{aligned} \tag{36}$$

Now, we show that $\text{Codim}_K(V_K, L_K) = d_y$, if $K = R$ or $K = \mathbb{C}$. Let V_K^* be the subspace of $L_K \times P(K^n)$, $P(K^n)$ the $n-1$ projective space, of all couples (L, l), such that $Cl = 0$, $Al \subset l$. Then, it is an algebraic submanifold of $L_K \times P(K^n)$, of codimension $n + d_y - 1$. If $\Pi : L_K \times P(K^n) \to L_K$ denotes the projection on the first factor, then, $\Pi(V_K^*) = V_K$. So, $\mathrm{Codim}_K(V_K, L_K) \geq n + d_y - 1 - (n-1) = d_y$. Let $\overset{\circ}{V}_K$ be the subset of all (C, A, l) in V_K^*, such that l is the eigenspace of A corresponding to a simple eigenvalue, and the only eigenvectors annihilated by C are those belonging to l. The set $\overset{\circ}{V}_K$ is open in V_K^*, and the restriction of Π to it is a diffeomorphism onto its image. So $\mathrm{Codim}_K(V_K, L_K) = d_y$ for $K = R$ or $\mathbb{C}$. Then, (36) implies that

$$\mathrm{Codim}_K(U_K, L_K) = d_y,$$

for $K = R$ or $\mathbb{C}$.

7.2. Lemmas

If V is a vector space, $Sym^a(V)$ denotes the space of symmetric tensors of degree a on V, which can be canonically identified with the homogeneous polynomials of degree a over V^*, dual of V. The symbol $\odot$ means **symmetric tensor product**.

To stress that the map Φ_k^Σ defined in Chapter 2, Section 3, depends only on the k-jet $j^k\Sigma$ of Σ, let us denote it by $\Phi^{j^k\Sigma}$ in this section. Let $(x, u^{(0)}, \hat{u}) \in X \times U \times R^{(k-1)d_u}$, and set $p = (x, u^{(0)})$. Let $f \in F$, such that $f(p) \neq 0$. Then, $j^k f(p) \in J^k F(p)$. Recall that $J^k H^0(p)$ is the subspace of $J^k H(p)$ formed by the $j^k h(p)$, such that h depends on x only.

Lemma 7.1.

(i) The mapping $\Theta_0 : J^k H^0 \to R^{kd_y}$, $j^k h \to \Phi^{j^k\Sigma}(x, u^{(0)}, \hat{u})$ *is linear and surjective,*

(ii) The mapping $\Theta_1 : J^k H^0 \to R^{kd_y} \otimes T^*X$, $j^k h \to T_X \Phi^{j^k\Sigma}(x, u^{(0)}, \hat{u})$ *is linear and surjective,*

(iii) For any $\Lambda \in T_x X$, $\Lambda \neq 0$, *the mapping* Θ_Λ,

$$\Theta_\Lambda : J^k H(p)\,(resp.\, J^k H^0(x)) \to R^{kd_y}, \left(p = \left(x, u^{(0)}\right)\right),$$

$$j^k h(p)\,(resp.\, j^k h(x)) \to T_X \Phi^{j^k\Sigma}\left(x, u^{(0)}, \hat{u}; \Lambda\right),$$

is linear and surjective.

Remark 7.1. It is not true that the mapping: $J^k H^0 \to J^1(X \times U, R^{kd_y})$, $j^k h(x) \to j^1 \Phi^{j^k\Sigma}(p, \hat{u})$, is surjective.

Proof. Let f be a representative of $j^k f(p)$, with $p = (x, u^{(0)})$. Take a coordinate system $(O, x_1, \ldots, x_n)$ for X at x. Then, if f^k denotes the kth dynamical extension of f:

$$(L_{f^k})^r h(p, \hat{u}) = d_X^r h(x; f(p)^{\odot r}) + \sum_{a=0}^{r-1} d_X^a h(x; R_{a,r}(j^k f(p), \hat{u})).$$

The expression $R_{a,r}$ is a universal polynomial mapping: $J^k f(p) \times R^{(k-1)d_u} \to Sym^a(T_x X)$. This gives immediately (i) because $\Theta_0(h) = (\ldots, (L_{f^k})^r h(p, \hat{u}), \ldots)$. Let $e_1, \ldots, e_n$ be a basis of $T_x X$ such that $f(p) = e_1$. Then,

$$d_X (L_{f^k})^r h(p, \hat{u})(e_i) = d_X^{r+1} h(x; e_i \odot e_1^{\odot r}) + \sum_{a=0}^{r} d_X^a h(x; Q_{a,r}(j^{r-a} f(p), \hat{u}, e_i)),$$

where $Q_{a,r} : J^{r-a} F(p) \times R^{(k-1)d_u} \to Sym^a(T_x X)$ *is* a universal mapping.

Because the elements $e_i \odot e_1^{\odot r}$ are linearly independent in $Sym^{r+1}(T_x X)$, the surjectivity of Θ_1 follows. Number (iii) can be proven in a similar way. ■

Set $G \subset J^k F \underline{\times_X} TX \times R^{(k-1)d_u}$, $(j^k f(p), \Lambda, \hat{u}) \in G$ if:

(i) $f(p) \neq 0$ (again, $p = (x, u^{(0)})$),
(ii) $\Lambda \neq 0$.

Let $\mu : J^k H \times_{X \times U} G$ (resp. $J^k H^0 \times_{X \times U} G) \to R^{kd_y}$, $\mu(j^k h(p), j^k f(p), \Lambda, \hat{u}) = T_X \Phi_k^{\Sigma}(p, \hat{u})\Lambda$, where $\Sigma = (f, h)$.

Lemma 7.2. *μ is a submersion.*

Proof. For any fixed $(j^k f(p), \Lambda, \hat{u}) \in G$, the mapping $j^k h(p)$ (resp. $j^k h(x_0)) \to \mu(j^k h(p), j^k f(p), \Lambda, \hat{u})$ (resp. $\mu(j^k h(x_0), j^k f(p), \Lambda, \hat{u}))$ is the linear mapping Θ_Λ of Lemma 7.1 above. By (iii) in that lemma, this linear mapping is surjective. Hence, μ is a submersion. ■

Lemma 7.3.

1. *Let $(j^k f(y), \Lambda, \hat{u})$ be an element of $J^k f \underline{\times_X} TX \times R^{(k-1)d_u}$ such that (i) $f(y) = 0$, (ii) $u^{(1)} = \ldots = u^{(\rho-1)} = 0$ $(\rho \geq 1)$, and (iii) the vectors $T_X f(y)^i \Lambda$, $0 \leq i \leq \rho$ are linearly independent, and (iv) $d_U f(y; u^{(\rho)}) \neq 0$. Then, with $y = (x_0, u^{(0)})$ and $\Sigma = (f, h)$, the linear mapping*

$$\Theta_\Lambda : J^k H(y) \to R^{kd_y} \ (\textit{resp. } \Theta_\Lambda : J^k H^0(x_0) \to R^{kd_y}),$$

$$j^k h(y) \to T_X \Phi^{j^k \Sigma}(y, \hat{u}; \Lambda) \ (\textit{resp. } j^k h(x_0)) \to T_X \Phi^{j^k \Sigma}(y, \hat{u}; \Lambda)),$$

is surjective.

2. *For $0 \leq N < \rho$:*

$$d_X (L_{f^k})^N h(y, \hat{u}; \Lambda) = d_X h(y; (T_X f(y))^N \Lambda), \ (\textit{resp. } d_X h(x_0; (T_X f(y))^N \Lambda)),$$

for $N = \rho$:

$$d_X(L_{f^k})^\rho h(y, \hat{u}; \Lambda) = d_X h(y; (T_X f(y))^\rho \Lambda) + d_X d_U h\big(y; \Lambda \otimes u^{(\rho)}\big),$$
$$(\textit{resp. } d_X h(x_0; (T_X f(y))^\rho \Lambda)).$$

Proof. In the following considerations, h has to be taken in H or in H^0 as a subspace of H by considering a function $h \in H^0$ as a function of $(x, u^{(0)})$ independent of $u^{(0)}$. This has no effect on our formulas other than simplifying them without cancelling the important terms. All the terms $d_X^p d_U^q h$ with $q > 0$ should be set equal to zero. If $h \in H^0$, when we write $h(y)$ or $d_X^r h(y; \ldots)$ for $y = (x_0, u^{(0)})$, we mean $h(x_0)$ or $d_X^r h(x_0; \ldots)$.

To prove 1, we have to compute Θ_Λ, at least partially. It is not easy to compute the Lie derivatives $d_X(L_{f^k})^r h$, so we have to proceed differently. We use the "flow" of f^k and computations with formal power series.

Let $(O, x^1, \ldots, x^n)$ be a coordinate system of X at x_0. There is an open set O_1, a number $\varepsilon > 0$ and a smooth mapping $\varphi :]-\varepsilon, \varepsilon[\times O_1 \to O, (t, x) \to \varphi(t, x)$, such that: (1) $x_0 \in O_1 \subset O$; (2) $\varphi(0, x) = x$ for $x \in O_1$; and (3) $\frac{d}{dt}\varphi(t, x) = f(\varphi(t, x), V(t))$ for all $(t, x) \in]-\varepsilon, \varepsilon[\times O_1$, where $V(t)$ is the polynomial mapping: $V(t) = \frac{1}{\rho!}u^{(\rho)}t^\rho + \ldots + \frac{1}{(k-1)!}u^{(k-1)}t^{k-1}$. For later purposes, set $v_i = \frac{1}{i!}u^{(i)}$. Hence, $V(t) = v_\rho t^\rho + \ldots + v_{k-1}t^{k-1}$. To compute $d_X(L_{f^k})^N h$, we use the following simple observations: let $u = (u^{(0)}, \hat{u})$. If $h \in H$ (resp. H^0):

1. $(L_{f^k})^N h(x, u) = \frac{\partial^N}{(\partial t)^N} h(\varphi(t, x), V(t))_{|t=0} = h_N(x, u)$, say,
2. $d_X(L_{f^k})^N h(x, u; \Lambda) = d_X h_N(x, u; \Lambda)$ for $\Lambda \in T_x X \approx R^n$ using the coordinate system. (Note that $h_0 = h$).

Let $\varphi_N(x, u) = \frac{1}{N!}\frac{\partial^N}{(\partial t)^N}\varphi(t, x)_{|t=0}$ for $x \in O_1$. Condition (2) on φ implies that $\varphi_0(x, u) = x$, for $x \in O_1$.

We denote by $\hat{\varphi}$ the Taylor series of φ at $t = 0$:

$$\hat{\varphi} = \sum_{r=0}^{\infty} t^r \varphi_r,$$

by $\hat{h}$ the Taylor series of $t \to h(\varphi(t, x), V(t))$ at $t = 0$:

$$\hat{h} = \sum_{r=0}^{\infty} t^r h_r.$$

$$\hat{h} = \sum_{p,q \geq 0} \frac{1}{p!}\frac{1}{q!} d_X^p d_U^q h(\hat{\varphi}, \ldots, \hat{\varphi}, V, \ldots, V), \ (p\ \varphi\text{'s and } q\ V\text{'s}). \quad (37)$$

Let us compute h_N and $d_X h_N$ for a multi-index $\alpha \in \mathcal{N}^l$, $|\alpha| = \alpha_1 + \ldots + \alpha_l$. For $x \in O_1$:

$$h_N(x,u) = \sum \left\{ \frac{1}{p!}\frac{1}{q!} d_X^p d_U^q h\big(x, u^{(0)}; \varphi_\alpha(x,u) \otimes v_\beta\big) \big| p \geq 0, \right.$$

$$\left. q \geq 0, \alpha \in \mathcal{N}^p, \beta \in \mathcal{N}^q, |\alpha| + |\beta| = N \right\},$$

where $\varphi_\alpha(x,u)$ (resp. v_β) will denote the symmetric tensor product $\varphi_{\alpha_1}(x,u) \odot \ldots \odot \varphi_{\alpha_p}(x,u)$ (resp. $v_{\beta_1} \odot \ldots \odot v_{\beta_q}$) of the vectors $\varphi_{\alpha_1}(x,u), \ldots, \varphi_{\alpha_p}(x,u) \in R^n$ (resp. $v_{\beta_1}, \ldots, v_{\beta_q} \in R^{d_u}$).

For any $x \in O_1$, any $\Lambda \in T_x X \approx R^n$:

$$d_X h_N(x,u;\Lambda) = I_N + II_N, \tag{38}$$

$$I_N = \sum \left\{ \frac{1}{p!}\frac{1}{q!} d_X^{p+1} d_U^q h\big(x, u^{(0)}; \Lambda \otimes \varphi_\alpha(x,u) \otimes v_\beta\big) \big| p \geq 0, \right.$$

$$\left. q \geq 0, \alpha \in \mathcal{N}^p, \beta \in \mathcal{N}^q, |\alpha| + |\beta| = N \right\}, \tag{39}$$

$$II_N = \sum \left\{ \frac{1}{p!}\frac{1}{q!} d_X^p d_U^q h\big(x, u^{(0)}; \varphi_{\alpha_1} \odot \ldots \odot d_X \varphi_{\alpha_i}(x,u;\Lambda) \odot \ldots \right.$$

$$\left. \odot\, \varphi_{\alpha_p}(x,u) \otimes v_\beta)\big) | p \geq 1, q \geq 0, 1 \leq i \leq p, \alpha \in \mathcal{N}^p, \beta \in \mathcal{N}^q, |\alpha| + |\beta| = N \right\}.$$

It is easy to see that

$$II_N = \sum \left\{ \frac{1}{(p-1)!}\frac{1}{q!} d_X^p d_U^q h\big(x, u^{(0)}; d_X \varphi_r(x,u;\Lambda) \otimes \varphi_\gamma(x,u) \otimes v_\beta\big) \right.$$

$$\left. | p \geq 1, q \geq 0, 1 \leq r, \gamma \in \mathcal{N}^{p-1}, \beta \in \mathcal{N}^q, r + |\gamma| + |\beta| = N \right\}. \tag{40}$$

Before proceeding further, we have to determine the φ_N and $d_X \varphi_N$ for small N. The claim is that

$$\varphi_N(x_0,u) = 0, 0 \leq N \leq \rho, \varphi_{\rho+1}(x_0,u)$$

$$= \frac{1}{\rho+1} d_U f\left(x_0, u^{(0)}; \frac{1}{\rho!} u^{(\rho)} \right). \tag{41}$$

$$d_X \varphi_N(x_0,u;\Lambda) = \frac{1}{N!} T_X f\big(x_0, u^{(0)}\big)^N \Lambda, 0 \leq N \leq \rho. \tag{42}$$

To see this, write the equation $\frac{d}{dt}\varphi(t,x) = f(\varphi(t,x), V(t))$ as a formal power series in t:

$$\sum_{N=1}^{\infty} N\varphi_N(x,u)t^{N-1} = \sum_{p,q\geq 0} \frac{1}{p!q!} d_X^p d_U^q f\big(x, u^{(0)}; \hat{\varphi}(t,x)^{\odot p} \otimes V(t)^{\odot q}\big). \tag{43}$$

Hence,

$$N\varphi_N(x,u) = \sum \left\{ \frac{1}{p!q!} d_X^p d_U^q f\big(x, u^{(0)}; \varphi_\alpha(x,u) \otimes v_\beta\big) \big| p \geq 0, q \geq 0, \alpha \in \mathcal{N}^p, \beta \in \mathcal{N}^q, |\alpha| + |\beta| = N-1 \right\}.$$

Because $v_i = 0$ for $i < \rho$, $|\beta| \geq q\rho$. This shows that if $N \leq \rho$:

$$\varphi_1(x,u) = f\big(x, u^{(0)}\big),$$

$$N\varphi_N(x,u^{(0)}) = \sum \left\{ \frac{1}{p!} d_X^p f\big(x, u^{(0)}; \varphi_\alpha(x,u)\big) \big| \alpha \in \mathcal{N}^p, |\alpha| = N-1 \right\}. \tag{44}$$

Because $f(x_0, u^{(0)}) = 0$, an easy induction on N shows that $\varphi_N(x_0, u) = 0$ if $1 \leq N \leq \rho$. In case $N = \rho + 1$, no terms with $p \geq 1$ can appear, because for any $\alpha \in N^p$, with $|\alpha| \leq \rho$, $\alpha_i \leq \rho$ for all i, $1 \leq i \leq p$. Hence,

$$(\rho+1)\varphi_{\rho+1}(x_0,u) = d_U f\left(x_0, u^{(0)}; \frac{1}{\rho!}u^{(\rho)}\right), \quad \left(\text{Note that } v_\rho = \frac{1}{\rho!}u^{(\rho)}\right).$$

Deriving (44), we get for $\Lambda \in T_x X \approx R^n$:

$$N d_X \varphi_N(x,u,\Lambda) = III_N + IV_N, \tag{45}$$

$$III_N = \sum \left\{ \frac{1}{p!} d_X^{p+1} f\big(x, u^{(0)}; \Lambda \otimes \varphi_\alpha(x,u)\big) \big| \alpha \in \mathcal{N}^p, |\alpha| = N-1 \right\},$$

$$IV_N = \sum \left\{ \frac{1}{(p-1)!} d_X^p f\big(x, u^{(0)}; d_X\varphi_r(x,u;\Lambda) \otimes \varphi_\gamma(x,u)\big) \big| \gamma \in \mathcal{N}^{p-1}, r \geq 1, |\gamma| + r = N-1 \right\}.$$

Because $\varphi_N(x_0,u) = 0$ if $N \leq \rho$, for $x = x_0$, this last formula reduces to: $N d_X \varphi_N(x_0, u; \Lambda) = d_X f(x_0, u^{(0)}; d_X \varphi_{N-1}(x_0, u; \Lambda))$. An easy induction gives: $d_X \varphi_N(x_0, u; \Lambda) = \frac{1}{N!} d_X f(x_0, u^{(0)})^N \Lambda$, for $0 \leq N \leq \rho$.

Let us go back to $d_X h_N$. In the light of Formulas (41) and (42), the maximum value of p appearing in a nonzero term in $I_{N_{|x=x_0}}$ is $m = [\frac{N}{\rho+1}]$, integer part of $\frac{N}{\rho+1}$ because $p(\rho+1) \leq |\alpha| \leq N$. In a nonzero term in $I_{N_{|x=x_0}}$ with $p = m$, $q\rho \leq |\beta| \leq N - m(\rho+1) \leq \rho$. Hence, $q = 0$ or $q = 1$. Finally, we get the following:

If $N - m(\rho+1) < \rho$,

$$I_{N_{|x=x_0}} = \frac{1}{m!}\sum\{d_X^{m+1}h(x_0, u^{(0)}; \Lambda \otimes \varphi_\alpha(x_0, u^{(0)}))|\alpha \in \mathcal{N}^m, |\alpha| = N\}$$
$$+\sum\left\{\frac{1}{p!q!}d_X^{p+1}d_U^q h(x_0, u^{(0)}; \Lambda \otimes \varphi_\alpha(x_0, u^{(0)}) \otimes v_\beta)|p \leq m-1,\right.$$
$$\left.\alpha \in \mathcal{N}^p, \beta \in \mathcal{N}^q, |\alpha| + |\beta| = N\right\},$$

if $N - m(\rho+1) = \rho$,

$$I_{N_{|x=x_0}} = \frac{1}{m!}\sum\{d_X^{m+1}h(x_0, u^{(0)}; \Lambda \otimes \varphi_\alpha(x_0, u^{(0)}))|\alpha \in \mathcal{N}^m, |\alpha| = N\}$$
$$+\frac{1}{m!}d_X^{m+1}d_U h(x_0, u^{(0)}; \Lambda \otimes \varphi_{\rho+1}(x_0, u^{(0)})^{\odot m} \otimes v_\rho)$$
$$+\sum\left\{\frac{1}{p!q!}d_X^{p+1}d_U^q h(x_0, u^{(0)}; \Lambda \otimes \varphi_\alpha(x_0, u^{(0)}) \otimes v_\beta)|p \leq m-1,\right.$$
$$\left.\alpha \in \mathcal{N}^p, \beta \in \mathcal{N}^q, |\alpha| + |\beta| = N\right\}.$$

The maximum value of p in a nonzero term $II_{N_{|x=x_0}}$ is $m+1$, because $(p-1)(\rho+1) \leq |\gamma| \leq N$. In a nonzero term in $II_{N_{|x=x_0}}$ with $p = m+1$, $q\rho \leq |\beta| \leq N - |\gamma| \leq N - m(\rho+1) \leq \rho$. If $N - m(\rho+1) < \rho$ then $q = 0$. If $N - m(\rho+1) = \rho$, $q\rho \leq |\beta| \leq N - |\gamma| - r \leq \rho - 1$ because $r \geq 1$. We get, setting $N - m(\rho+1) = \varepsilon$:

$$II_{N_{|x=x_0}} = A_N + B_N,$$

$$A_N = \frac{1}{m!}\sum\{d_X^{m+1}h(x_0, u^{(0)}; d_X\varphi_r(x_0, u; \Lambda) \otimes \varphi_\gamma(x_0, u))|$$
$$1 \leq r \leq \varepsilon, \gamma \in \mathcal{N}^m, |\gamma| + r = N\}$$

$$B_N = \sum\left\{\frac{1}{(p-1)!q!}d_X^p d_U^q h(x_0, u^{(0)}; d_X\varphi_r(x_0, u; \Lambda) \otimes \varphi_\gamma(x_0, u) \otimes v_\delta)\right|$$
$$\left.1 \leq r \leq \varepsilon, p \leq m, \gamma \in \mathcal{N}^{p-1}, \delta \in \mathcal{N}^q, r + |\gamma| + |\delta| = N\right\}.$$

Putting $I_{N_{|x=x_0}}$, $II_{N_{|x=x_0}}$ together and using Formulas (41) and (42): Let N be an integer between 0 and $k-1$, let $m = [\frac{N}{\rho+1}]$ and $\varepsilon = N - m(\rho+1)$, $0 \le \varepsilon \le \rho$.

If $\varepsilon < \rho$ and $\Lambda \in T_{x_0}X \approx R^n$,

$$d_X h_N\big(x_0, u^{(0)}; \Lambda\big) = d_X^{m+1} h\big(x_0, u^{(0)}; \Psi_N\big) + \sum \big\{d_X^p d_U^q h\big(x_0, u^{(0)}; \Psi_{N,p,q}\big)\big| 1 \le p \le m, q \le N\big\}, \tag{46}$$

if $\varepsilon = \rho$ and $\Lambda \in T_{x_0}X \approx R^n$,

$$\begin{aligned} d_X h_N\big(x_0, u^{(0)}; \Lambda\big) &= d_X^{m+1} h\big(x_0, u^{(0)}; \Psi_N\big) \\ &\quad + d_X^{m+1} d_U h\left(x_0, u^{(0)}; \Lambda \otimes \frac{1}{m!}\varphi_{\rho+1}(x_0, u)^{\odot m} \otimes v_\rho\right) \\ &\quad + \sum \big\{d_X^p d_U^q h\big(x_0, u^{(0)}; \Psi_{N,p,q}\big)\big| 1 \le p \le m, q \le N\big\}, \end{aligned} \tag{47}$$

where

$$\Psi_N = \Psi_N(j^1 f(y), \Lambda, \hat{u}) = \sum_{r=0}^{\varepsilon} \frac{1}{r!} T_X f(y)^r \Lambda \otimes \frac{1}{m!} U_m(\varepsilon - r), \tag{48}$$

$$U_m(i) = \sum \{\varphi_\gamma(x_0, u) | \gamma \in \mathcal{N}^m, |\gamma| = m(\rho+1) + i\}. \tag{49}$$

In particular

$$U_m(0) = \varphi_{\rho+1}(x_0, u)^{\odot m} = \frac{1}{(\rho+1)^m}\left[d_U f\left(x_0, u^{(0)}; \frac{1}{\rho!}u^{(\rho)}\right)\right]^{\odot m}, \tag{50}$$

$$\begin{aligned} \Psi_{N,p,q} &= \Psi_{N,p,q}(j^N f(y), \Lambda, \hat{u}) \\ &= \frac{1}{(p-1)!q!} \sum \big\{d_X \varphi_r(x_0, u; \Lambda) \otimes \varphi_\gamma(x_0, u) \otimes v_\delta | \\ &\qquad r \ge 0, \gamma \in \mathcal{N}^{p-1}, \delta \in \mathcal{N}^q, r + |\gamma| + |\delta| = N\big\}. \end{aligned} \tag{51}$$

Clearly, Ψ_N is a polynomial mapping:

$$J^k S(y) \times T_{x_0}X \times R^{(k-1)d_u} \to Sym^{m+1}(T_{x_0}X), \text{ linear in } \Lambda,$$

and $\Psi_{N,p,q}$ is a polynomial mapping:

$$J^k S(y) \times T_{x_0}X \times R^{(k-1)d_u} \to Sym^p(T_{x_0}X) \otimes Sym^q(R^{d_u}).$$

In view of the expressions (48), (49), and (50) of the Ψ_N and the assumptions (iii) and (iv) of Part 1 of the lemma, the elements $\{\Psi_N(j^1 f(y), \Lambda, \hat{u}) \mid m(\rho+1) \leq N \leq m(\rho+1)+\rho\}$ in $Sym^{m+1}(T_{x_0}X)$ are linearly independent. Because $\Theta_\Lambda(j^k h(y)) = (d_X h_0(x_0, u; \Lambda), d_X h_1(x_0, u; \Lambda), \ldots, d_X h_{k-1}(x_0, u; \Lambda))$, Formulas (46) and (47) show that Θ_Λ is surjective. Part 2 of the lemma follows from Formulas (46) and (47): If $m = 0$, $U_0(i) = 0$ for $i > 1$, $U_0(0) = 1$, and $\Psi_{N,p,q} = 0$ since $1 \leq p \leq m = 0$, which is impossible. ∎

5
Singular State-Output Mappings

In the two previous chapters, all *initial – state → output – trajectory* mappings are regular in some sense: either the system has a uniform canonical flag, and it is also, at least locally, strongly differentially observable (see Exercise 4.1 of Chapter 3), or, in the case where $d_y > d_u$, systems are generically strongly differentially observable of some order. In both cases, the *initial – state → output – trajectory* mapping is an immersion in some sense, and as a consequence, the systems have the phase variable property.

It can happen that the *initial – state → output – trajectory* mapping is not an immersion, but that nevertheless, the system possesses the phase variable property of some order.

It is interesting to study these singular situations because, for observation or output stabilization, **only the phase variable property matters**, as will be clear in the next chapters. This study is the purpose of this chapter.

The uncontrolled case is very different from the controlled one. We will show that, **in the uncontrolled analytic case**, a reasonable assumption is that the map Φ_N^Σ **is a finite mapping** for some N. Unfortunately, in this case, there is no C^∞ version of our results.

In both cases (controlled and uncontrolled), the first step of the study is **local** (at the level of germs of systems). Afterward, assuming observability (injectivity), the phase variable representations can be glued together using a partition of unity.

On the other hand, **in the controlled case, we do not need the analyticity assumption**. However, for the sake of simplicity of the exposition, we let it stand.

1. Assumptions and Definitions

Here, we consider only **analytic** systems, of the following form:

$$(\Sigma)\ \frac{dx}{dt} = f(x, u),\ \ y = h(x, u), \text{(controlled case), or,}$$

$$(\Sigma)\ \frac{dx}{dt} = f(x),\ \ y = h(x), \text{(uncontrolled case).}$$

As a first step, we will consider germs of such systems at a point $(x_0, u^{(0)}) \in X \times U$ (controlled case), or $x_0 \in X$ (uncontrolled case). In this chapter, U is not assumed to be compact. In most cases, for global considerations, we will consider that $U = R^{d_u}$.

1.1. Notations

Again in this section the value $u = u^{(0)}$ of the control plays a role different from the higher order derivatives $u^{(i)}$, $i \geq 1$. Hence, we introduce the following notations.

Given a $N-$jet $\underline{f}_{N+1} = (f^{(0)}, f^{(1)}, \ldots, f^{(N)})$ of a curve f in a Euclidean space R^m, we will write

$$\tilde{f}_N = \left(f^{(1)}, \ldots, f^{(N)}\right),$$
$$\underline{f}_{N+1} = \left(f^{(0)}, \tilde{f}_N\right). \tag{52}$$

We use this notation in the case of infinite jets ($N = \infty$), and we drop the subscript $N = \infty$ ($\underline{f}_\infty = \underline{f}$, $\tilde{f}_\infty = \tilde{f}$).

Let us define the restricted mappings $\Phi^{\Sigma}_{N,\tilde{u}_{N-1}} : X \times U \to R^{Nd_y}$ and $S\Phi^{\Sigma}_{N,\tilde{u}_{N-1}} : X \times U \to R^{Nd_y} \times R^{d_u}$,

$$\Phi^{\Sigma}_{N,\tilde{u}_{N-1}}\left(x_0, u^{(0)}\right) = \Phi^{\Sigma}_N\left(x_0, u^{(0)}, \tilde{u}_{N-1}\right),$$
$$S\Phi^{\Sigma}_{N,\tilde{u}_{N-1}}\left(x_0, u^{(0)}\right) = \left(\Phi^{\Sigma}_N\left(x_0, u^{(0)}, \tilde{u}_{N-1}\right), u^{(0)}\right). \tag{53}$$

Let O_{x_0} be the ring of germs of analytic functions at $x_0 \in R^d$. Let $\Re$ be a subring of O_{x_0}, $x_0 \in R^d$. For $u_0 \in R^p$, $\Re\{u; u_0\}$ will denote the ring of germs at $(x_0, u_0) \in R^d \times R^p$ of analytic mappings of the form $G(u, \varphi_1(x), \ldots, \varphi_r(x))$, for G analytic at $(u_0, \varphi_1(x_0), \ldots, \varphi_r(x_0))$, and for any finite subset $\{\varphi_1, \ldots, \varphi_r\} \subset \Re$. If $\Re$ is an analytic algebra (in the sense of [43], for instance), then $\Re\{u; u_0\}$ is also an analytic algebra.

1.2. Rings of Functions

We have to consider several rings of (germs of) analytic functions attached to the germ of a system Σ at a point. There are different definitions for the controlled and uncontrolled case. Let us fix a point $x_0 \in X$, and an infinite jet $(u_0^{(0)}, \widetilde{u_0}) = \underline{u_0}$.

Let us define the rings $\Re_N(x_0)$, or $\Re_N(x_0, \underline{u_0}_N)$, $\hat{\Re}_N(x_0, \underline{u_0}_N)$.

1. **In the uncontrolled case:**

$$\Re_N(x_0) = \left(\Phi^{\Sigma}_N\right)^*(O_{y_0}), \tag{54}$$

the pull back by Φ_N^Σ of the ring O_{y_0} of germs of analytic real valued functions $\varphi(y, \tilde{y}_{N-1})$ at the point $y_0 = \Phi_N^\Sigma(x_0)$, i.e.,

$$\Re_N = \left\{ G \circ \Phi_N^\Sigma(x) \middle| G \text{ is an analytic germ at } y_0 = \Phi_N^\Sigma(x_0) \right\}. \qquad (55)$$

2. **In the controlled case:**

$$\Re_N(x_0, \underline{u_0}_N) = \left(S\Phi^\Sigma_{N,\tilde{u}_{0,N-1}}\right)^*(O_{y_0}),$$
$$\hat{\Re}_N(x_0, \underline{u_0}_N) = \left(S\Phi^\Sigma_N\right)^*(O_{z_0}),$$

where $y_0 = S\Phi^\Sigma_{N,\tilde{u}_{0,N-1}}(x_0, u_0^{(0)})$ and $z_0 = S\Phi^\Sigma_N(x_0, u_0^{(0)}, \widetilde{u}_{0N-1})$.

If there is no ambiguity about the choice of x_0, $(u_0^{(0)}, \widetilde{u}_{0N-1}) = \underline{u_0}_N$, we will write $\Re_N, \Re_N, \hat{\Re}_N$ in place of $\Re_N(x_0)$, $\Re_N(x_0,\underline{u_0}_N)$, $\hat{\Re}_N(x_0,\underline{u_0}_N)$. For each N, $\hat{\Re}_N$ can be canonically identified to a subring of $\hat{\Re}_{N+1} : \hat{\Re}_N \subset \hat{\Re}_{N+1}$.

In both the controlled and the uncontrolled case, $\Re_N$ and $\hat{\Re}_N$ are Noetherian rings. They form increasing sequences:

$$\begin{aligned} &\ldots \subset \Re_N \subset \Re_{N+1} \subset \ldots \subset O_{x_0} \text{ or } O_{(x_0,u_0^{(0)})}, \\ &\ldots \subset \hat{\Re}_N \subset \hat{\Re}_{N+1} \subset \ldots \end{aligned} \qquad (56)$$

In the controlled case, $\breve{\Re}$ will denote the ring of germs of analytic mappings of the form $G(u, \varphi_1, \ldots, \varphi_p)$ at the point $(x_0, u_0^{(0)})$, for any positive integer p and for functions φ_i of the form

$$\varphi_i = L_{f_u}^{k_1}\left(\partial_{j_1}\right)^{s_1} L_{f_u}^{k_2}\left(\partial_{j_2}\right)^{s_2} \ldots\ldots L_{f_u}^{k_r}\left(\partial_{j_r}\right)^{s_r} h, \; k_l, s_l \geq 0, \qquad (57)$$

(see Formula (15) in the definition of Ξ^Σ, Chapter 2). Recall that ∂_j denotes the derivation with respect to the jth control variable.

Obviously, $\breve{\Re}$ is closed under the action of the derivations L_{f_u} and ∂_j, $1 \leq j \leq d_u$. Also,

$$\Re_N \subset \breve{\Re} \text{ for all } N. \qquad (58)$$

$\breve{\Re}_N$ will denote the ring of germs of analytic mappings of the form $G(u, \varphi_1, \ldots, \varphi_p)$ at the point $(x_0, u_0^{(0)})$, for all functions φ_i of the form (57) above, with $\sum k_i + \sum s_i \leq N - 1$.

Remark 1.1.

1. The ring $\hat{\Re}_N\{u^{(N)}, u_0^{(N)}\}$ is exactly the ring of germs at a point of (analytic) elements of the rings $\Re_N^\sharp$, defined in Chapter 2, Section 3,

2. The ring $\check{\mathfrak{R}}$ is just the ring of analytic germs at $(x_0, u_0^{(0)})$ generated by the germs of the elements of the space Ξ^{Σ}, plus the control variables, (Ξ^{Σ} has been defined in Section 5 of Chapter 2).

2. The Ascending Chain Property

Definition 2.1. A germ of analytic system Σ (at the point x_0 or at the point $(x_0, u_0^{(0)}, \widetilde{u_0}))$ satisfies the "ascending chain property of order N," denoted by $ACP(N)$, if:

Uncontrolled case: $\mathfrak{R}_j = \mathfrak{R}_N$ for $j \geq N$,
Controlled case: $\hat{\mathfrak{R}}_{j+1} = \hat{\mathfrak{R}}_j\{u^{(j)}; u_0^{(j)}\}$ for $j \geq N$.

Convention: For simplicity, we will say that a vector function belongs to $\mathfrak{R}_N$ or $\hat{\mathfrak{R}}_N$ if each of its components does.

The next two lemmas show the relation between the ascending chain property $ACP(N)$ and the phase-variable property $PH(N)$. The phase-variable property $PH(N)$ has been defined in Chapter 2 for systems. For **germs of** analytic systems, the definition is similar and left to the reader.

Lemma 2.1. *Σ satisfies the $ACP(N)$ at some point iff $\mathfrak{R}_{N+1} = \mathfrak{R}_N$, (resp. $\hat{\mathfrak{R}}_{N+1} = \hat{\mathfrak{R}}_N\{u^{(N)}; u_0^{(N)}\}$) in the uncontrolled (resp. controlled) case.*

Lemma 2.2. *Each of the following two conditions is necessary and sufficient for Σ to satisfy the $ACP(N)$:*

(i) $y^{(N)} = \Psi_N(y^{(0)}, \tilde{y}_{N-1}, u^{(0)}, \tilde{u}_N)$ for some analytic function Ψ_N (locally defined in a neighborhood of $(S\Phi_N^{\Sigma}(x_0, u_0^{(0)}, \widetilde{u_0}_{N-1}), u_0^{(N)})$;
(ii) $y^{(j)} = \Psi_j(y^{(0)}, \tilde{y}_{N-1}, u^{(0)}, \tilde{u}_j)$ for some analytic function Ψ_j locally defined and for all $j \geq N$.

In the uncontrolled case, the conditions (i) and (ii) of Lemma 2.2 give: (i) $y^{(N)} = \Psi_N(y^{(0)}, \tilde{y}_{N-1})$, and (ii) $y^{(j)} = \Psi_j(y^{(0)}, \tilde{y}_{N-1})$.

Remark 2.1. A priori condition (i) is necessary for Σ to satisfy the $ACP(N)$, and condition (ii) is sufficient.

Remark 2.2. The second (resp. first) condition of Lemma 2.2 is equivalent to the phase variable property $PH(j)$ of any order $j \geq N$ (resp. the phase variable property $PH(N)$ of order N).

Let us prove these lemmas, considering the controlled case only.

Proof of Lemma 2.1. Assume that Σ satisfies the $ACP(N)$ at $(x_0, u_0^{(0)}, \widetilde{u}_{0N-1})$. Then, by definition, $\hat{\Re}_{N+1} = \hat{\Re}_N\{u^{(N)}; u_0^{(N)}\}$.

For the converse, assume that $\hat{\Re}_{N+1} = \hat{\Re}_N\{u^{(N)}; u_0^{(N)}\}$. By definition,

$$\hat{\Re}_{N+2} \supset \hat{\Re}_{N+1}\{u^{(N+1)}; u_0^{(N+1)}\}.$$

Let $\varphi \in \hat{\Re}_{N+2}$. Then, for some analytic G, we have:

$$\varphi = G\big(y^{(0)}, \tilde{y}_{N+1}, u^{(0)}, \tilde{u}_{N+1}\big).$$

For $y_i^{(N)}$ denoting the ith component of $y^{(N)}$, $1 \leq i \leq d_y$, we have $y_i^{(N)} \in \hat{\Re}_{N+1} = \hat{\Re}_N\{u^{(N)}; u_0^{(N)}\}$. Hence,

$$y_i^{(N)} = \Psi^i\big(y^{(0)}, \tilde{y}_{N-1}, u^{(0)}, \tilde{u}_N\big) \text{ for some } \Psi^i.$$

Therefore,

$$\begin{aligned} y_i^{(N+1)} &= \dot{y}_i^{(N)} = d\Psi^i\big(y^{(0)}, \tilde{y}_{N-1}, u^{(0)}, \tilde{u}_N\big).\frac{d}{dt}_{|t=0}\big(y^{(0)}, \tilde{y}_{N-1}, u^{(0)}, \tilde{u}_N\big) \\ &= H^i(y^{(0)}, \tilde{y}_N, u^{(0)}, \tilde{u}_{N+1}). \end{aligned}$$

This implies that

$$\begin{aligned} \varphi &= G\big(y^{(0)}, \tilde{y}_N, H\big(y^{(0)}, \tilde{y}_N, u^{(0)}, \tilde{u}_{N+1}\big), u^{(0)}, \tilde{u}_{N+1}\big) \\ &\in \hat{\Re}_{N+1}\big\{u^{(N+1)}; u_0^{(N+1)}\big\}. \end{aligned}$$

Hence, $\hat{\Re}_{N+2} = \hat{\Re}_{N+1}\{u^{(N+1)}; u_0^{(N+1)}\}$. ■

Remark 2.3. It is easy to check that the condition $\hat{\Re}_{i+1} = \hat{\Re}_i\{u^{(i)}; u_0^{(i)}\}$ for $i \geq N$ is equivalent to $\hat{\Re}_{N+k} = \hat{\Re}_N\{u^{(N)}, \ldots, u^{(N+k-1)}; u_0^{(N)}, \ldots, u_0^{(N+k-1)}\}$ for $k \geq 1$.

Proof of Lemma 2.2. If Σ satisfies the $ACP(N)$ at $(x_0, u_0^{(0)}, \widetilde{u}_0)$, then, by definition, $y_i^{(N)} = \Psi_{N,i}(y^{(0)}, \tilde{y}_{N-1}, u^{(0)}, \tilde{u}_N)$. Conversely, if $y^{(N)} = \Psi_N(y^{(0)}, \tilde{y}_{N-1}, u^{(0)}, \tilde{u}_N)$, by definition again, if $\varphi \in \hat{\Re}_{N+1}$, then

$$\begin{aligned} \varphi &= G\big(y^{(0)}, \tilde{y}_N, u^{(0)}, \tilde{u}_N\big) \\ &= G\big(y^{(0)}, \tilde{y}_{N-1}, \Psi_N\big(y^{(0)}, \tilde{y}_{N-1}, u^{(0)}, \tilde{u}_N\big), u^{(0)}, \tilde{u}_N\big) \in \hat{\Re}_N\big\{u^{(N)}; u_0^{(N)}\big\}, \end{aligned}$$

and

$$\hat{\Re}_{N+1} \subset \hat{\Re}_N\big\{u^{(N)}; u_0^{(N)}\big\}.$$

By Lemma 2.1, Σ satisfies the $ACP(N)$. This proves (i). Applying (i), an obvious induction gives (ii). ■

3. The Key Lemma

3.1. Finite Germs

Here, we prove a lemma about the ascending chain property, which will be used later on. Let $(f_j, j > 0)$ be a sequence of analytic germs: $(X, x_0) \to (Y, f_j(x_0))$. X, Y are analytic manifolds. As was done previously in a particular case, we can associate a sequence of local rings $\Re_j$ to the sequence (f_j) in the following way: We denote by $\Phi_j : X \to Y^j$ the map $\Phi_j(x) = (f_1(x), \ldots, f_j(x))$, and by $\Re_j$:

$$\Re_j = (\Phi_j)^*\big(O_{\Phi_j(x_0)}\big),$$

the pull back by the map Φ_j of the ring $O_{\Phi_j(x_0)}$ of germs of analytic maps at the point $\Phi_j(x_0)$. Clearly, again we have

$$\ldots \subset \Re_j \subset \Re_{j+1} \subset \ldots \subset O_{x_0}.$$

Definition 3.1. We say that the sequence (f_j) satisfies the $ACP(N)$ at x_0, if $\Re_j = \Re_N$ for $j \geq N$.

Definition 3.2. **(of finite multiplicity)**. $F : X \to Y$ has finite multiplicity at x_0 if $O_{x_0}/[F^*(\mathrm{m}(O_{y_0})).O_{x_0}]$ has finite dimension as a real vector space. Here $\mathrm{m}(O_{y_0})$ is the ideal of germs of analytic functions at (Y, y_0), $y_0 = F(x_0)$, which are zero at y_0.

The dimension is the multiplicity.

There is a simple and convenient criterion for a germ to be of finite multiplicity:

F has finite multiplicity at x_0 iff there is an integer $r > 0$ such that:

$$\big[\mathrm{m}(O_{x_0})\big]^r \subset F^*\big(\mathrm{m}(O_{y_0})\big).O_{x_0}. \tag{59}$$

Therefore, to check that F has finite multiplicity at $x_0 = 0$, $(F : R^n \to Y)$, it is sufficient to check that $x_i^{r_i}$ belongs to $F^*(\mathrm{m}(O_{y_0})).O_{x_0}$ for some positive integers r_i, $i = 1, \ldots, n$.

Let $\Phi_N : (X, x_0) \to Y^N$, $\Phi_N = (f_1, \ldots, f_N)$, where the $f_i : (X, x_0) \to Y$ are germs of mappings at x_0. A "prolongation of Φ_N" is an arbitrary sequence $(\hat{f}_j)$ of germs of mappings, $\hat{f}_j : (X, x_0) \to Y$, such that $\hat{f}_j = f_j$ for $j \leq N$.

3.2. The Lemma

We have the following key lemma:

Lemma 3.1. (see, [32]). *The following properties are equivalent:*

(i) All prolongations of Φ_N satisfy the $ACP(k)$ for some $k \geq N$, (k depends on the prolongation),
(ii) Φ_N has finite multiplicity.

Proof. (ii)$\Longrightarrow$ (i): We are going to use the following crucial property ($\mathcal{P}$) (for the proof, see [43], Chapter 1): If a germ Φ_N has finite multiplicity at x_0, then O_{x_0} is a finitely generated $\Re_N$ module. This last fact is also expressed by saying that Φ_N **is a finite germ at** x_0.

For any given prolongation of Φ_N, the associated sequence of local rings $\Re_j$ satisfies: $\Re_N \subset \Re_{N+1} \subset \ldots \subset O_{x_0}$, and each of the $\Re_j$, $j \geq N$, is a $\Re_N$ submodule of O_{x_0}.

As the image of a Noetherian ring by a morphism of rings, $\Re_N$ is a Noetherian ring. Any finitely generated module over a Noetherian ring is Noetherian module over this ring (see [53], p. 158, Theorem 18). Therefore, O_{x_0} is a Noetherian $\Re_N$ module. Hence, any increasing sequence of $\Re_N$ submodules is stationary. This shows that the sequence $\Re_j$ is stationary, which means that the prolongation satisfies the $ACP(k)$ for some k.

(i)$\Rightarrow$ (ii). This part of the proof uses the same trick as the proof of Nakayama's lemma.

We consider $\Phi_N = (f_1, \ldots, f_N)$. We can assume that $Y = R$ and $f_1(x_0) = \ldots = f_N(x_0) = 0$. A prolongation $(\hat{f}_j)$ being given, let $\mathrm{m}(\Re_j) \subset \Re_j$ be the maximal ideal of $\Re_j$. The ideal generated by $\mathrm{m}(\Re_j)$ in O_{x_0} is denoted by $\mathrm{m}(\Re_j).O_{x_0}$. We will first construct a prolongation $(\hat{f}_j)$ such that $\mathrm{m}(\Re_r) = \mathrm{m}(\Re_N).O_{x_0}$ for some r. We chose $\hat{f}_{N+1} \in \mathrm{m}(\Re_N).O_{x_0}$, but $\hat{f}_{N+1} \notin \mathrm{m}(\Re_N)$, $\hat{f}_{N+2} \in \mathrm{m}(\Re_N).O_{x_0}$, but $\hat{f}_{N+2} \notin \mathrm{m}(\Re_{N+1})$, and so on. For such a sequence, $\mathrm{m}(\Re_r) \neq \mathrm{m}(\Re_{r+1})$ if $r \geq N$, hence, $\Re_r \subset \Re_{r+1}$ but $\Re_r \neq \Re_{r+1}$ for $r \geq N$. This contradicts the assumption (i) of the lemma. Therefore, for some r, $\mathrm{m}(\Re_r) = \mathrm{m}(\Re_N).O_{x_0}$.

Now, $\mathrm{m}(\Re_N).O_{x_0}$ is obviously a finitely generated $\Re_r$ module (it is generated by $\hat{f}_1, \ldots, \hat{f}_r$). Also

$$x_1 \hat{f}_i \in \mathrm{m}(\Re_N).O_{x_0}, \quad 1 \leq i \leq r,$$

hence,

$$x_1 \hat{f}_i = \sum_{k=1}^{r} a_{i,k} \hat{f}_k, \quad \text{where } a_{i,k} \in \Re_r.$$

Therefore,

$$(x_1 Id - A)\Phi_r = 0,$$

where A is the matrix formed by the $a_{i,k}$.

It follows that the determinant $\Delta = \det(x_1 Id - A) = 0$. Expanding this determinant, we get

$$x_1^r + \sum_{i<r} \alpha_i x_1^i = 0, \text{ with } \alpha_i \in \mathfrak{R}_r.$$

For each i, we have $\alpha_i = \beta_i + \gamma_i$, where β_i is a constant, γ_i is a nonunit of $\mathfrak{R}_r$. We have

$$x_1^r + \sum_{i<r} \beta_i x_1^i = -\sum_{i<r} \gamma_i x_1^i \in \mathrm{m}(\mathfrak{R}_N).O_{x_0}.$$

Evaluating this formula at $x = x_0 = 0$ (which we can assume), we get that $\beta_0 = 0$. If all the β_i, $i < r$, are zero, then, $x_1^r \in \mathrm{m}(\mathfrak{R}_N).O_{x_0}$. If one is nonzero, then, for some $p < r$, $x_1^{r-p}(x_1^p + \cdots + \delta_2 x_1 + \delta_1) \in \mathrm{m}(\mathfrak{R}_N).O_{x_0}$, with δ_i constant and $\delta_1 \neq 0$. Hence $(x_1^p + \cdots + \delta_2 x_1 + \delta_1)$ is a unit in O_{x_0}. This shows that for some $l > 0$, $x_1^l \in \mathrm{m}(\mathfrak{R}_N).O_{x_0}$. The same is true for all the variables x_i, $1 \leq i \leq n$. By the criterion, Φ_N has finite multiplicity. ∎

3.3. Why Analyticity?

If one looks at the classical proof of Property $(\mathcal{P})$ above, one notices that it essentially follows from the **Weierstrass Preparation Theorem** (see [43]). If we assume that our mappings are C^∞, then Lemma 3.1 is false. A counterexample is provided below.

The step *(i)* $\Longrightarrow$ *(ii)* of the lemma is still valid in the C^∞ case but the step *(ii)* $\Longrightarrow$ *(i)* is not. Using the Thom–Malgrange theory (see [49], page 189, Corollary 3.3), it is still true that O_{x_0} is a finitely generated $\mathfrak{R}_N$ module, but the proof of *(ii)* $\Longrightarrow$ *(i)* breaks down because $\mathfrak{R}_N$ is not in general Noetherian. We shall construct a smooth map Φ_2 on R^2 with finite multiplicity, and a smooth prolongation of it, which does not satisfy the $ACP(k)$ for any k. This will imply that $\mathfrak{R}_N$, in this counterexample, is not Noetherian.

Here, we show a smooth map Φ_2 on R^2, with finite multiplicity, and a smooth prolongation that does not satisfy the $ACP(k)$ for any k.

3.4. Counterexample

Let W be the "Weierstrass manifold," $W = \{(x_0, x_1, t) | t^2 + x_1 t + x_0 = 0\} \subset R^3$, and $\Pi : W \to R^2$, $(x_0, x_1, t) \to (x_0, x_1)$. Certainly, W is a smooth manifold. Set $f_0 = x_0$, $f_1 = x_1$ and $f_n = g_n(x_0, x_1)t$, with the sequence g_n constructed as follows.

Consider on R^2 the polar coordinates (r, θ), and the vector field X:

$$X = -r\frac{\partial}{\partial r} + \frac{\partial}{\partial \theta}.$$

We can construct some **spiraloid** disjoint subsets S_n of R^2 as follows: We pick an interval $I_1 = [a_1, b_1] \subset R_+^*$, I_1 small enough for $S_1 \cap \{x_1 = 0\} \neq \{x_1 = 0\}$, where S_1 is the union set of all trajectories of X passing through the points $(x^0, 0)$, $x^0 \in I_1$. Now, we choose a second interval $I_2 = [a_2, b_2]$, with $0 < a_2 < b_2 < a_1$ and $I_2 \cap S_1 \cap \{x_1 = 0\} = \varnothing$, and construct the set S_2 as the union set of all the trajectories of X through $(I_2, 0)$. Iterating the construction, we get S_n. We chose g_n in such a way that its support is $cl(S_n)$, the closure of S_n, and $g_n > 0$ on $\text{Int}(S_n)$. This is possible because the complement of this set is closed, and since, given any closed set, there exists a C^∞ function having this set as zero set.

The multiplicity of $F = (f_0, f_1)$, $F : W \to R^2$ is finite at $(0, 0, 0)$, (it is 2).

We show that the sequence f_n does not satisfy the $ACP(k)$ for any k. For this, we work in an arbitrary small ball B centered at $(0, 0, 0)$ in W. We assume that $f_{n+1} = \Psi(f_0, f_1, \ldots, f_n)$ on B for some smooth Ψ. By construction, if $p = \Pi p' \in Int(S_{n+1})$, then $f_{n+1}(p') = \Psi(f_0(p'), f_1(p'), 0, \ldots 0)$. Let D be the discriminant set of W (i.e., $D = \{(x_0, x_1, t) | x_0 = \frac{1}{4}x_1^2\}$). We consider $c = (c_0, c_1)$, $c \in \Pi B \cap Int(S_{n+1}) \cap \Pi D$, $c' \in B \cap \Pi^{-1}(c)$, and a sequence (p'_k) in $\Pi^{-1}(Int(S_{n+1}))\backslash D$ such that $\lim_{k\to\infty} p'_k = c'$, and we set $p_k = \Pi p'_k$. By definition, we have

$$g_{n+1}(x_0, x_1)t = \Psi(x_0, x_1) \text{ on } Int(S_{n+1}).$$

Differentiating, we get

$$\frac{\partial g_{n+1}}{\partial x_0}(x_0, x_1)t + g_{n+1}(x_0, x_1)\frac{\partial t}{\partial x_0} = \frac{\partial \Psi}{\partial x_0}(x_0, x_1),$$

which should hold at p_k,

$$\frac{\partial g_{n+1}}{\partial x_0}(p_k)t(p_k) + g_{n+1}(p_k)\frac{\partial t}{\partial x_0}(p_k) = \frac{\partial \Psi}{\partial x_0}(p_k).$$

Taking the limit when $k \to \infty$, we get

$$\frac{\partial g_{n+1}}{\partial x_0}(c)t(c) + g_{n+1}(c)\frac{\partial t}{\partial x_0}(c) = \frac{\partial \Psi}{\partial x_0}(c),$$

where $g_{n+1}(c)$ is different from zero, and $t(c) = -\frac{c_1}{2}$. Hence, $\frac{\partial t}{\partial x_0}(c)$ is well defined. However, $t^2 + x_1 t + x_0 = 0$ implies $\frac{\partial t}{\partial x_0} = -\frac{1}{2t+x_1}$. At c, $x_1 = c_1$, $t(c) = -\frac{c_1}{2}$, and $2t + x_1 = 0$. This is a contradiction.

Hence, despite the fact that the multiplicity is finite, the equality

$$f_{n+1} = \Psi(f_0, \ldots, f_n),$$

never holds.

3.5. Consequences of the Key Lemma

The main consequence of Lemma 3.1 is the following theorem.

Theorem 3.2. *Let Σ be an uncontrolled system. Let $x_0 \in X$ be fixed. If for some k, Φ_k^Σ has finite multiplicity at x_0, then Σ satisfies the $ACP(N)$ at x_0 for some $N \geq k$, and by Lemma 2.2, $y^{(N)} = \Psi_N(y^{(0)}, \tilde{y}_{N-1})$ for some analytic function Ψ_N defined in a neighborhood of $\Phi_N^\Sigma(x_0)$.*

Example 3.1. $X = R$, $y = h(x)$, $\dot{x} = f(x)$, x_0 is arbitrary, h is nonconstant. In that case, of course, the notion of multiplicity is equivalent to the usual natural notion of multiplicity of a smooth function of a single variable. The multiplicity is always finite because h is nonconstant, and hence for some N, we have (locally): $y^{(N)} = \Psi_N(y^{(0)}, \tilde{y}_{N-1})$.

Example 3.2. $X = R^2$, $y = x_1$, $\dot{x}_1 = x_2^3$, $\dot{x}_2 = f(x_1, x_2)$, $x_0 = (0, 0)$. This system is observable, and by our criterion, the multiplicity is finite. Hence, for some N, we have also: $y^{(N)} = \Psi_N(y^{(0)}, \tilde{y}_{N-1})$.

Of course, it can happen that $\Phi_N = (f_1, \ldots, f_N)$ does not have finite multiplicity, but some particular prolongations $(\hat{f}_r)$ satisfy the $ACP(k)$ for some k. (Just take the prolongation by the zero sequence for instance.)

For uncontrolled systems, there are many other interesting examples where the $ACP(N)$ holds for some N, but the multiplicity is not finite. A case where the $ACP(N)$ holds everytime is the following:

Exercise 3.1. (Linear systems observed polynomially). $X = R^n$, $y = p(x)$ is a polynomial, $\dot{x} = Ax$ is a linear vector field. Show that the $ACP(N)$ holds for some N (compare with Exercise 4.8, Chapter 3).

Exercise 3.2. $(\Sigma) : X = R^2$, $y = x_1(x_1^2 + x_2^2)$, $A = \begin{bmatrix} 0 & 1 \\ -1 & 0 \end{bmatrix}$.

1. Show that Σ is observable.
   By the previous exercise, the $ACP(N)$ holds: $y^{(2)} = -y^{(0)}$.
2. Show that the multiplicity is infinite. (For hints, see [30].)

The following important theorem will be proven later on as a consequence of more general results in the controlled case. It is a **globalization** of Theorem 3.2.

Theorem 3.3. *(Globalization of the $ACP(N)$ in the uncontrolled case). Assume that X is compact, Σ is observable, and for each $x_0 \in X$, there is an N (depending on x_0, it could be), such that Φ_N^Σ has finite multiplicity at x_0. Then there is a k and a C^∞ function φ, defined and compactly supported on R^{kd_y}, such that*

$$y^{(k)}(x) = \varphi\big(y^{(0)}(x), \tilde{y}_{k-1}(x)\big), \text{ for all } x \in X,$$

that is, Σ satisfies $PH(k)$, the phase variable property of order k, globally on X.

4. The $ACP(N)$ in the Controlled Case

As stated in Theorem 3.2, the local $ACP(N)$ holds in the uncontrolled case, as soon as there is a k such that Φ_k^Σ has finite multiplicity at the point under consideration. As we shall see, in the controlled case, the situation is not so clear cut.

Here, as we have said, **the C^ω assumption could be relaxed**. Let us keep it for the clarity of exposition. But, in Chapter 7, we will use the results in the case where the systems are C^∞.

The main local result is the following.

Theorem 4.1. *A point x_0 and an infinite jet $\underline{u_0} = (u_0^{(0)}, \widetilde{u_0})$ are fixed. Σ satisfies the $ACP(N)$ iff $\check{\mathfrak{R}} = \mathfrak{R}_N$. Moreover, in that case, $\mathfrak{R}_N = \mathfrak{R}_{N+1}$, $\mathfrak{R}_N \subset \hat{\mathfrak{R}}_N$.*

Proof. To start, let us prove that the $ACP(N)$ implies $\check{\mathfrak{R}} = \mathfrak{R}_N$. This comes from the following lemma (Lemma 4.2): $h \in \mathfrak{R}_N \cap \hat{\mathfrak{R}}_N$, $\mathfrak{R}_N \cap \hat{\mathfrak{R}}_N$ is stable by L_{f_u} and ∂u_j, $1 \leq j \leq d_u$. Therefore, by definition, $\check{\mathfrak{R}} \subset \mathfrak{R}_N$. But, $\mathfrak{R}_N \subset \check{\mathfrak{R}}$ by definition. Hence, $\check{\mathfrak{R}} = \mathfrak{R}_N$. Also, obviously, in that case, $\mathfrak{R}_N = \mathfrak{R}_{N+1}$.

Let us show that $\check{\mathfrak{R}} = \mathfrak{R}_N$ implies that $\mathfrak{R}_N \subset \hat{\mathfrak{R}}_N$. We denote temporarily by $\bar{Y}$ the vector $\bar{Y}(x, u^{(0)}, \tilde{u}_{N-1}) = (y^{(0)}, y^{(1)}, \ldots, y^{(N-1)}) = \underline{y_N}$, by $Z(x, u)$ the vector $Z = (y(x, u^{(0)}), \; y^{(1)}(x, u^{(0)}, \widetilde{u_{01}}), \ldots, y^{(N-1)}(x, u^{(0)}, \widetilde{u_{0N-1}}))$, and by $Z'(x, u^{(0)})$ the vector of all expressions $L_{f_u}^{k_1}(\partial_{j_1})^{s_1} L_{f_u}^{k_2}(\partial_{j_2})^{s_2} \ldots . L_{f_u}^{k_r}(\partial_{j_r})^{s_r} h$, where $\sum k_i + \sum s_i \leq N-1$ (the generators of $\check{\mathfrak{R}}_N$).

An obvious computation shows that

$$\bar{Y}\big(x, u^{(0)}, \tilde{u}_{N-1}\big) = Z\big(x, u^{(0)}\big) + \varphi\big(Z'\big(x, u^{(0)}\big), \tilde{u}_{N-1}\big), \tag{60}$$

where φ is some polynomial mapping and $\varphi(Z', \widetilde{u_{0N-1}}) = 0$ for all Z'.

Just to show how it works, the two first components of the above equality are, in the single input case,

$$y^{(0)}(x, u^{(0)}) = h(x, u^{(0)}),$$
$$y^{(1)}(x, u^{(0)}, u^{(1)}) = L_{f_u}h + \partial_u h(u_0)^{(1)} + \partial_u h(u^{(1)} - (u_0)^{(1)}).$$

Now, assuming that $\check{\mathfrak{R}} = \mathfrak{R}_N$, $\mathfrak{R}_N$ is stable by ∂_j and L_{f_u}. Therefore, h being in $\mathfrak{R}_N$, all the components of $Z'(x, u^{(0)})$ are in $\mathfrak{R}_N$, hence, $Z' = \Psi(Z, u^{(0)})$ for a certain Ψ. Therefore, the relation (60) can be rewritten as

$$\bar{Y}(x, u^{(0)}, \tilde{u}_{N-1}) = Z(x, u^{(0)}) + \varphi(\Psi(Z, u^{(0)}), \tilde{u}_{N-1}), \tag{61}$$

and the differential of $\varphi(\Psi(Z, u^{(0)}), \tilde{u}_{N-1})$ with respect to Z at $\widetilde{u_0}_{N-1}$ is zero. Hence, the map H,

$$H(Z, u^{(0)}, \tilde{u}_{N-1}) = (Z + \varphi(\Psi(Z, u^{(0)}), \tilde{u}_{N-1}), u^{(0)}, \widetilde{u}_{N-1})$$

is a diffeomorphism from an open neighborhood U^0 of $(Z(x_0, (u_0)^{(0)}), (u_0)^{(0)}, \widetilde{u_0}_{N-1})$, onto an open neighborhood V^0 of $(\bar{Y}(x_0, (u_0)^{(0)}, \widetilde{u_0}_{N-1}), (u_0)^{(0)}, \widetilde{u_0}_{N-1})$. In particular, $Z(x, u^{(0)}) = G(\bar{Y}(x, u^{(0)}, \tilde{u}_{N-1}), u^{(0)}, \tilde{u}_{N-1})$, for all $(x, u^{(0)}, \tilde{u}_{N-1}) \in W^0$, a certain neighborhood of $(x_0, (u_0)^{(0)}, \widetilde{u_0}_{N-1})$. Therefore, $Z(x, u^{(0)}) \in \hat{\mathfrak{R}}_N$, and $\varphi(Z, u^{(0)}) \in \hat{\mathfrak{R}}_N$ for all $\varphi : \mathfrak{R}_N \subset \hat{\mathfrak{R}}_N$.

To complete the proof, let us show that $\check{\mathfrak{R}} \subset \hat{\mathfrak{R}}_N$ implies the $ACP(N)$. If $\check{\mathfrak{R}} \subset \hat{\mathfrak{R}}_N$, then $\check{\mathfrak{R}}\{\tilde{u}_{N-1}; \widetilde{u_0}_{N-1}\} \subset \hat{\mathfrak{R}}_N$. But, by definition, $\hat{\mathfrak{R}}_N \subset \check{\mathfrak{R}}\{\tilde{u}_{N-1}; \widetilde{u_0}_{N-1}\}$, hence $\check{\mathfrak{R}}\{\tilde{u}_{N-1}; \widetilde{u_0}_{N-1}\} = \hat{\mathfrak{R}}_N$. Otherwise, also,

$$\hat{\mathfrak{R}}_{N+1} \subset \check{\mathfrak{R}}\{\tilde{u}_N; \widetilde{u_0}_N\} = \hat{\mathfrak{R}}_N\{u^{(N)}; (u_0)^{(N)}\}.$$

Therefore, $\hat{\mathfrak{R}}_{N+1} = \hat{\mathfrak{R}}_N\{u^{(N)}; (u_0)^{(N)}\}$, and, by definition, the $ACP(N)$ holds. ■

Lemma 4.2. *If the $ACP(N)$ holds, then:*

(i) $\partial_j(\mathfrak{R}_N \cap \hat{\mathfrak{R}}_N) \subset \mathfrak{R}_N \cap \hat{\mathfrak{R}}_N$, $j = 1, \ldots, d_u$,
(ii) $L_{f_u}(\mathfrak{R}_N \cap \hat{\mathfrak{R}}_N) \subset \mathfrak{R}_N \cap \hat{\mathfrak{R}}_N$.

Proof. First, if $\varphi \in \hat{\mathfrak{R}}_N$, $\frac{d^k\varphi}{dt^k} \in \hat{\mathfrak{R}}_{N+k}$, (where $\frac{d}{dt}$ is the operator defined by the dynamics of Σ in the obvious way): By definition,

$$\varphi = G(y^{(0)}, \tilde{y}_{N-1}, u^{(0)}, \tilde{u}_{N-1}),$$
$$\frac{d\varphi}{dt} = dG(y^{(0)}, \tilde{y}_{N-1}, u^{(0)}, \tilde{u}_{N-1}). \frac{d}{dt}(y^{(0)}, \tilde{y}_{N-1}, u^{(0)}, \tilde{u}_{N-1})$$
$$= H(y^{(0)}, \tilde{y}_N, u^{(0)}, \tilde{u}_N) \in \hat{\mathfrak{R}}_{N+1}.$$

An obvious induction gives the result.

Second, we know (Remark 2.3 after the proof of Lemma 2.1) that

$$\hat{\mathfrak{R}}_{N+k} = \hat{\mathfrak{R}}_N\{u^{(N)}, \ldots, u^{(N+k-1)}; u_0^{(N)}, \ldots, u_0^{(N+k-1)}\}.$$

Hence, if $\varphi \in \mathfrak{R}_N \cap \hat{\mathfrak{R}}_N$, then

$$\frac{d^N\varphi}{dt^N} \in \hat{\mathfrak{R}}_{2N} = \hat{\mathfrak{R}}_N\{u^{(N)}, \ldots, u^{(2N-1)}; u_0^{(N)}, \ldots, u_0^{(2N-1)}\}.$$

However, an obvious computation shows that

$$\frac{d^N\varphi}{dt^N} = \varphi_1(\tilde{u}_{N-1}) + \sum_{j=1}^{d_u} u_j^{(N)} \partial_j(\varphi),$$

because $\varphi \in \mathfrak{R}_N$. Therefore, since neither φ_1 nor $\partial_j(\varphi)$ depend on $u^{(N)}$, $\partial_j(\varphi) \in \hat{\mathfrak{R}}_N$.

Now, $\frac{d\varphi}{dt} = L_{f_u}(\varphi) + \sum_{j=1}^{d_u} u_j^{(1)}\partial_j(\varphi) \in \hat{\mathfrak{R}}_{N+1} = \hat{\mathfrak{R}}_N\{u^{(N)}; u_0^{(N)}\}$. Hence $L_{f_u}(\varphi) + \sum_{j=1}^{d_u} u_j^{(1)}\partial_j(\varphi) \in \hat{\mathfrak{R}}_N$. Because $\partial_j(\varphi) \in \hat{\mathfrak{R}}_N$, this implies that $L_{f_u}(\varphi) \in \hat{\mathfrak{R}}_N$.

$L_{f_u}(\varphi)$ and $\partial_j(\varphi)$ all belong to $\hat{\mathfrak{R}}_N$ and do not depend on $u^{(i)}$, $i \geq 1$. Hence, $L_{f_u}(\varphi)$ and $\partial_j(\varphi)$ all belong to $\mathfrak{R}_N$ by definition. ■

Remark 4.1.

(i) If the $ACP(N)$ is true at $(x_0, u_0^{(0)}, \widetilde{u_0})$ for some $\widetilde{u_0}$, then,

$$\check{\mathfrak{R}}_N = \check{\mathfrak{R}}_{N+1} \text{ at } (x_0, u_0^{(0)}),$$

(ii) This condition $\check{\mathfrak{R}}_N = \check{\mathfrak{R}}_{N+1}$ is implied by the fact that $\check{\mathfrak{R}}_{N_0}$ has finite multiplicity for some N_0 (in the sense that the map with components the generators of $\check{\mathfrak{R}}_{N_0}$, given by Formula (57), has finite multiplicity).

Exercise 4.1. $(\Sigma) : X = R^2$, $U = R$, $y = x_1$,

$$\dot{x}_1 = x_2^3 - x_1,$$
$$\dot{x}_2 = x_2^8 + x_2^4 u.$$

We work at $x_0 = (0, 0)$, and $u_0^{(0)} = 0$.

Show that:

1. Σ is observable.
2. $\check{\mathfrak{R}} = \mathfrak{R}_4 = \{G(x_1, x_2^3, x_2^{10}, x_2^{17}, u)\}$,
   so that the $ACP(4)$ holds (in fact it holds as soon as $(x_0)_2 = 0$, and the $ACP(2)$ holds everywhere else).

3. Show that actually, $y^{(4)}$ can be written as a polynomial:

$$y^{(4)} = P\big(y^{(0)}, y^{(1)}, y^{(2)}, y^{(3)}, u^{(0)}, u^{(1)}, u^{(2)}\big),$$

and compute P.

5. Globalization

5.1. Preliminaries

We assume that Σ is given, and we fix a compact subset K of X. We also assume that Σ is differentially observable of order N. We denote by $\Phi_N^{\Sigma,K}$ the following mapping:

Uncontrolled case: $\Phi_N^{\Sigma,K}$ is the restriction of Φ_N^{Σ} to K,
$\Phi_N^{\Sigma,K} : K \to \Phi_N^{\Sigma,K}(K) \subset R^{Nd_y}$,

Controlled case: $\Phi_N^{\Sigma,K}$ is the restriction of $S\Phi_N^{\Sigma}$ to $K \times U \times R^{(N-1)d_u}$,
$\Phi_N^{\Sigma,K} : K \times U \times R^{(N-1)d_u} \to \Phi_N^{\Sigma,K}(K \times U \times R^{(N-1)d_u}) \subset R^{Nd_y} \times R^{Nd_u}$.

Lemma 5.1. *$\Phi_N^{\Sigma,K}$ is a homeomorphism onto its image, which is closed.*

Proof. $\Phi_N^{\Sigma,K}$ is proper. Apply Proposition 2, p. 113 of [7]. ■

Comments

1. Lemma 5.1 shows that, as soon as Σ is differentially observable, x can be expressed on K as a continuous function φ_N^K of $(y^{(0)}, \tilde{y}_{N-1}, u^{(0)}, \tilde{u}_{N-1})$.
2. If $X = R^n$, then, by Urysohn's lemma, φ_N^K can be extended to a continuous function defined on all of $R^{Nd_y} \times R^{Nd_u}$, hence, x can be written as a continuous function, defined on all of $R^{Nd_y} \times R^{Nd_u}$:

$$x = \varphi_N^K\big(y^{(0)}, \tilde{y}_{N-1}, u^{(0)}, \tilde{u}_{N-1}\big).$$

3. If $X = R^n$, (or if X is not R^n but φ_N^K is globally defined and continuous on $R^{Nd_y} \times R^{Nd_u}$), the classical assumption that φ_N^K is smooth is equivalent to the **strong** differential observability assumption (in restriction to K). It is much stronger than the differential observability assumption made here. Of course, it implies (in the uncontrolled case) that the multiplicity is finite. Actually, the multiplicity is one.

It is the case in Chapters 3 and 4. In both chapters, strong differential observability holds (by assumption in Chapter 4, and as a consequence of uniform infinitesimal observability in Chapter 3).

Exercise 5.1. Consider the system Σ:

$$(\Sigma)\ \dot{x} = 1,\ y = \cos(x) + \cos(\alpha x),\ \ x \in R,$$

where α is an irrational number.

1. Show that Σ is observable.
2. Show that the observation space Θ^{Σ} of Σ is finite dimensional. Compare with Exercise 3.1. Therefore the $ACP(N)$ holds for some N.
3. Show that x cannot be expressed as a continuous function of $(y^{(0)}, \tilde{y}_M)$, whatever M.

Exercise 5.2. Show that, in Example 3.2 (uncontrolled):

1. Σ is differentially observable of order 2.
2. Depending on the choice of f, it can happen that Φ_N^{Σ} is an immersion for some N, or Φ_N^{Σ} is not an immersion for any N.

Exercise 5.3. Show that, in Exercise 4.1:

1. Σ is differentially observable of order 2.
2. $S\Phi_N^{\Sigma}$ is not an immersion for any N.

5.2. Main Result

The main result in this Section is the following Theorem:

Theorem 5.2. *Assume that Σ satisfies the $ACP(N)$ at each point, and is differentially observable of order N. Consider K, any fixed compact subset of X. Then, there exists a C^∞ function φ_N^K,* ***compactly supported w.r.t.*** $(y^{(0)}, \tilde{y}_{N-1})$*, such that*

$$y^{(N)} = \varphi_N^K\big(y^{(0)}, \tilde{y}_{N-1}, u^{(0)}, \tilde{u}_N\big), \tag{62}$$

for all $x \in K$, all $u, \tilde{u}_N$. That is, Σ satisfies $PH(N)$, the phase variable property of order N, in restriction to K.

Compactly supported w.r.t. $(y^{(0)}, \tilde{y}_{N-1})$ means that, for any K', a compact subset of $U \times R^{Nd_u}$, φ_N^K restricted to $R^{Nd_y} \times K'$, is compactly supported. (It is not equivalent that φ_N^K is compactly supported, for all fixed $u, \tilde{u}_N$.)

Proof. Let $Z = (x, u^{(0)}, \tilde{u}_N)$ denote a typical element of $X \times U \times R^{Nd_u}$, $Z = (Z_0, \tilde{u}_N)$, $Z_0 = (x, u^{(0)}) \in X \times U$. Also, set $S'\Phi_N^{\Sigma}(x, u, \tilde{u}_N) = (S\Phi_N^{\Sigma}(x, u, \tilde{u}_{N-1}), u^{(N)})$.

For each Z, there is a neighborhood U^Z of Z, and a map φ^Z defined on V^Z, V^Z open, relatively compact in $R^{Nd_y} \times R^{(N+1)d_u}$ such that

$$y^{(N)} = \varphi^Z\left(y^{(0)}, \tilde{y}_{N-1}, u^{(0)}, \tilde{u}_N\right) \text{ on } U^Z.$$

We consider $S'\Phi_N^\Sigma(U^Z) = V_0^Z$, $V_0^Z \cap V^Z = V_0^Z$, and

$$V^Z \cap S'\Phi_N^\Sigma(K \times U \times R^{Nd_u}) = S'\Phi_N^\Sigma(U^Z). \tag{63}$$

This is possible since $S'\Phi_N^\Sigma$ is a homeomorphism onto its image when restricted to $K \times U \times R^{Nd_u}$ by Lemma 5.1. We take U^Z, V^Z so small that the diameter of V^Z is less than 1 in $R^{Nd_y} \times R^{(N+1)d_u}$. We also consider $V^0 = (R^{Nd_y} \times R^{(N+1)d_u}) \backslash S'\Phi_N^\Sigma(K \times U \times R^{Nd_u})$. This set V^0 is open, and we set $\varphi^0 = 0$.

The family $\{V^Z | Z \in K \times U \times R^{Nd_u}\} \cup \{V^0\}$ forms an open covering of $R^{Nd_y} \times R^{(N+1)d_u}$. By paracompactness, we can find a locally finite refinement of this covering, denoted by $\{W^i | i \in I\}$. To each of these W^i, we associate φ^i as follows: W^i is contained in V^0, or in V^{Z_i} for some $Z_i \in K \times U \times R^{Nd_u}$. If W^i is contained in V^0, then we set $\varphi^i = \varphi^0 = 0$. If W^i is contained in V^{Z_i} (we select one), then we set $\varphi^i = \varphi^{Z_i}$. We chose a partition of unity $\{\chi^i\}$ subordinated to this locally finite open covering $\{W^i\}$, and we set,

$$\varphi = \sum_{i \in I} \chi^i \varphi^i.$$

Clearly, φ is a C^∞ function, compactly supported w.r.t. $(y^{(0)}, \tilde{y}_{N-1})$. It remains to prove that the equality (62) of the theorem holds.

Let $Z \in K \times U \times R^{Nd_u}$. Set $\omega = S'\Phi_N^\Sigma(Z)$. Then, ω belongs to a certain finite number of sets W^i, say, $W^1, \ldots, W^k$. All we have to prove is that $\varphi^i(\omega) = y^{(N)}(Z)$ for $i = 1, \ldots, k$. This follows immediately from the injectivity of $S'\Phi_N^\Sigma$ and the property (63) of the V^{Z_i} stated at the beginning of the proof. ■

5.3. Consequences

The following corollary will be also used in Chapter 7.

Corollary 5.3. *Theorem 5.2 is valid not only for $y^{(N)}$, but also for any function α in $\hat{\mathfrak{R}}_{N+1} = \hat{\mathfrak{R}}_N\{u^{(N)}; u_0^{(N)}\}$ (i.e., the germs of α at each point belong to $\hat{\mathfrak{R}}_{N+1}$). It is true also for any function α in $\hat{\mathfrak{R}}_N$, and in that case:*

$$\alpha = \varphi_N^K\left(y^{(0)}, \tilde{y}_{N-1}, u^{(0)}, \tilde{u}_{N-1}\right), \tag{64}$$

for all $x \in K$, all $(u^{(0)}, \tilde{u}_{N-1})$.

Proof. In the proof of Theorem 5.2, we only used the fact that α belongs to $\hat{\mathfrak{R}}_N\{u^{(N)}; u_0^{(N)}\}$ at each point. Moreover, if α depends only on $(x, u^{(0)}, \tilde{u}_{N-1})$, then, applying the result, we get that $\alpha(x, u^{(0)}, \tilde{u}_{N-1}) = \varphi_N^K(y^{(0)}, \tilde{y}_{N-1}, u^{(0)}, \tilde{u}_{N-1}, 0)$ for all $x \in K$, all $(u^{(0)}, \tilde{u}_{N-1})$. ■

Example 5.1. Example 3.2 and Exercise 4.1 (see also the Exercises 5.2 and 5.3) satisfy, in the uncontrolled and controlled cases, the assumptions of Theorem 5.2. In the case of Exercise 4.1, it has been already stated that Formula (62) holds globally, (for φ_N^K a certain polynomial).

Exercise 5.4. (Single output, $d_y = 1$). Let Σ be a system with uniform canonical flag. Remember that Σ satisfies the phase variable property $PH(n)$ in restriction to small neighborhoods of each point in X (Exercise 4.1, Chapter 3). Moreover, assume that Σ is differentially observable of some order N. Show that Σ satisfies the $PH(N)$ (in restriction to any compact subset K of X).

Now, as a consequence, we can give the proof of Theorem 3.3, in the uncontrolled case.

Proof of Theorem 3.3. Because Σ is observable, then the observation space $\Theta^\Sigma = span\{L_f^k h(x) | k \geq 0\}$ separates points (see Exercise 5.2, Chapter 2).

Define $V^m \subset X \times X$:

$$V^m = \{(x_1, x_2) | L_f^k h(x_1) = L_f^k h(x_2), 0 \leq k \leq m\}.$$

The sequence V^m is a decreasing sequence of analytic subsets of $X \times X$, which is compact, and $\cap_{m \geq 0} V^m = \Delta X$, the diagonal in $X \times X$. By (Corollary 1, p. 99 of [43]), V^m is a stationary sequence: for some N, $V^N = \Delta X$. Hence, Φ_N^Σ is injective and Σ is differentially observable of order N. Also, each $x \in X$ has an open neighborhood U^x such that the $ACP(N^x)$ holds on U^x. Extracting a finite open covering leads to an N' such that the $ACP(N')$ holds at all $x \in X$. Applying Theorem 5.2 for $N = \sup(N, N')$ gives the result. ■

6. The Controllable Case

Let us assume that Σ is controllable, in the usual weak sense of the transitivity of its Lie algebra (see the Appendix of Chapter 2). We will use the Theorems 5.1 and 5.2, of Chapter 2, which in this case allow us to conclude that the **trivial foliation Δ_Σ is regular**, and equal to the foliation $\bar{\Delta}_\Sigma$, as was already stated just after the proof of Theorem 5.2 in Chapter 2.

Assume that these foliations are not trivial (i.e., their dimension is strictly > 0). Then, as was also stated in Section 5 of Chapter 2, Σ cannot be observable, for any fixed input. Hence, Σ cannot be differentially observable.

The consequence in the (analytic) controlled case, **if Σ is controllable**, is that **this part of the theory is void**: assume that, as in the assumptions of Theorem 5.2, Σ satisfies the $ACP(N)$ and is differentially observable. Then, the "trivial foliation" has to be trivial, which implies that $\Xi^{\Sigma}(u)$ has full rank everywhere, hence, *rank*$(d_X\check{\mathfrak{R}}) = n$. By Theorem 4.1, the $ACP(N)$ holds iff $\check{\mathfrak{R}} = \mathfrak{R}_N$, and in that case, $\mathfrak{R}_N \subset \hat{\mathfrak{R}}_N$. Therefore, *rank*$(d_X\hat{\mathfrak{R}}_N) = n$, and $S\Phi_N^{\Sigma}$ is an immersion. In fact, we are back to the situation of Chapters 3 or 4, where Σ is strongly differentially observable.

Exercise 6.1. Study the controllability (in the weak sense) for the system of Exercise 4.1.

6

Observers: The High-Gain Construction

The subject of this chapter is **observers**. The purpose of an observer is to obtain information about the state of the system, from the observed data. In Section 1 of this chapter, we are going to discuss the concept of observers. The main ingredient of that concept is the notion of "estimation." There is no completely satisfactory definition of estimation. For that reason, we have to present several definitions of an observer, each having its domain of application. We shall explain these different definitions of observers and point out the relations between them.

In the remainder of the chapter, we shall construct explicitly several types of observers. The fundamental idea behind all of these constructions is to use the classical observers for linear systems and to kill the nonlinearities by an appropriate **time rescaling**.

The construction and its variations that we present provide explicit, efficient, and robust algorithms for state estimation. It is closely related to the results of the three previous chapters (3, 4, 5), and it applies in all the cases which were dealt with in these chapters.

Our observers can be used for several purposes:

1. State estimation in itself.
2. Dynamic output stabilization.

They will be used in Chapter 7 for purpose 2.

Defining an observer presents several problems. First, there is no good definition of a state observer when **the state space X is not compact**. A second difficulty is the **peak phenomenon**, explained below in the Section 1.2.2, for observers with arbitrary exponential decay.

Finally, let us point out that our construction of observers is related to nonlinear filtering theory. But this is beyond the scope of this book. A good reference for this relation is Reference [12].

In this chapter, we will make the following basic assumption: The system (Σ) is **differentially observable** of a certain order $N \geq 1$.

1. Definition of Observer Systems and Comments

An observer system is a system Σ_O, the inputs of which consist of the **observed data** of Σ, i.e., the inputs of Σ, their derivatives, and the outputs of Σ. The task of the observer is the estimation of the state of Σ.

Let us make a few remarks. The inputs are selected by the "operator" of the system. In particular, he can chose them differentiable, and then, their derivatives are known. On the other hand, it is hard to estimate the derivatives of the outputs from the knowledge of them. For this reason, we strictly avoid any use of these derivatives in our theory. Actually, the first problem we shall deal with will be to estimate the derivatives of the outputs.

Because Σ is differentially observable, **for sufficiently smooth controls**, estimating the state is equivalent to estimating the $N-1$ first derivatives of the outputs, $\tilde{y}_{N-1}$.

We denote by $U^{r,B}$ the set of inputs $u : [0, \infty[\rightarrow U = I^{d_u}$, such that u is $r-1$ times continuously differentiable, its rth derivative belongs to $L^{\infty}([0, \infty[; R^{d_u})$ and all the derivatives up to order r are bounded by $B > 0$. The set $U^{0,B}$ is just the subset of $L^{\infty}([0, \infty[; U)$ formed by the $u(.)$ that are bounded by B. (Here, $U = I^{d_u}$ is not necessarily compact: $I = R$ is possible). The norm $\|.\|$ is **the canonical Euclidean norm on R^{d_y} or on R^{Nd_y}**.

1.1. Output Observers

1.1.1. Definitions

We use again the notation $\tilde{u}_r = (u^{(1)}, \ldots, u^{(r)})$, introduced in Section 1 of Chapter 5.

Definition 1.1. An $U^{r,B}$ output observer of Σ, relative to Ω is a system $(\Sigma_{Oy}^{r,B,\Omega})$, $r \geq N$:

$$(\Sigma_{Oy}^{r,B,\Omega}) \begin{cases} \frac{dz}{dt} = F\big(z, u^{(0)}(t), \tilde{u}_r(t), y^{(0)}(t)\big), \\ \eta(t) = \mathcal{H}\big(z(t), u^{(0)}(t), \tilde{u}_r(t), y^{(0)}(t)\big), \end{cases} \tag{65}$$

on the d_Z dimensional manifold Z, where $\Omega \subset X$ is open; where η, the output, belongs to R^{Nd_y}; F and $\mathcal{H}$ are C^{∞}, and satisfy the following condition:

For all $u \in U^{r,B}$, for all $x_0 \in \Omega$, such that the corresponding semi trajectory of Σ, $x(t, x_0)$, is defined on $[0, +\infty[$ and stays in Ω, for all $z_0 \in Z$, the output $\eta(t)$ is well defined and,

$$\lim_{t \rightarrow +\infty} \|\eta(t) - \underline{y}_N(t)\| = 0. \tag{66}$$

Remark 1.1. We will be mostly interested in two cases: X is compact and $\Omega = X$, or X is noncompact but Ω is relatively compact.

Definition 1.2 below strengthens Definition 1.1.

Definition 1.2. An **exponential** $U^{r,B}$ output observer of Σ, relative to Ω, is a one parameter family of output observers of Σ, depending on the real parameter $\alpha > 0$, with state manifold Z, independent of α, which satisfies the following condition (67), strengthening (66):

For any $\bar{K}$, a compact subset of Z, for all $z_0 \in \bar{K}$, for all $x_0 \in \Omega$, for all $u \in U^{r,B}$:

$$\|\eta(t) - \underline{y}_N(t)\| \leq k(\alpha)e^{-\alpha t}\|\eta(0) - \underline{y}_N(0)\|, \tag{67}$$

as long as $x(t, x_0)$ stays in Ω, where $k : R_+ \to R_+$ is a function of polynomial growth, depending on $\bar{K}$ in general.

Such a one parameter family will be typically denoted by $(\Sigma_{Oye}^{r,B,\Omega})$.

Remark 1.2. The fact that $k(\alpha)$ has polynomial growth warrants that, if α is large enough, **the estimate can be made arbitrarily close to the real value in arbitrary short time.**

1.1.2. Comment

Let us assume that $(\Sigma_{Oy}^{r,B,\Omega})$ is an "output observer" and $r = N$. Because Σ is differentially observable of order N, we obtain an estimation $\hat{x}(t)$ of the state $x(t)$ of Σ as follows.

For Ω' open, $cl(\Omega) \subset \Omega'$, denote by $E(t, z_0)$ the set:

$$\begin{aligned}\big\{x^* \in cl(\Omega')\big| \; &\big\|\eta(t) - \Phi_N^\Sigma(x^*, u(t), \tilde{u}_{N-1}(t))\big\| \\ &= \inf_{x\in\Omega'} \big\|\eta(t) - \Phi_N^\Sigma(x, u(t), \tilde{u}_{N-1}(t))\big\|\big\}.\end{aligned} \tag{68}$$

If Ω is relatively compact, then Ω' can be taken relatively compact. In that case, this set $E(t, z_0)$ is not empty, and for any metric d on X, compatible with its topology, $\lim_{t\to+\infty} d(E(t, z_0), x(t)) = 0$. **If Ω is not relatively compact, this is not true.**

Exercise 1.1. In this situation, show that $\lim_{t\to+\infty} \#(E(t, z_0)) = 1$, if moreover Σ is **strongly** differentially observable (of order N), for a trajectory $x(t, x_0)$ that stays in Ω for all $t \geq 0$.

1.1.3. The Observability Distance

The following could be a way to overcome the problems linked to the noncompactness of X or Ω: One should try to construct a canonical distance over X, related to the observability properties. This canonical distance could then be used in the definition of observers. A trivial way to do this is to define the distance on X:

$$d_O(x,y) = \sup_{\|u^{(0)}\|, \|\tilde{u}_{N-1}\| \leq B} \left\| \Phi_N^\Sigma\left(x, u^{(0)}, \tilde{u}_{N-1}\right) - \Phi_N^\Sigma\left(y, u^{(0)}, \tilde{u}_{N-1}\right) \right\|. \tag{69}$$

Exercise 1.2. Show that (69) actually defines a distance over X.

Unfortunately, this distance is not compatible with the topology of X in general:

Exercise 1.3. Consider the system of Exercise 5.1 in Chapter 5 (uncontrolled case). Show that the observability distance is not compatible with the topology of X.

Moreover, this distance is not very natural because it depends on both B (the bound on the input and its derivatives) and on N (the degree of differential observability).

There is a special case where the situation is better: if X is compact, and if Σ is analytic, uncontrolled and just observable, then by the proof of Theorem 3.3 in the previous Chapter 5, there is an N such that Σ is differentially observable of order N. Taking the smallest such N, we get a canonical observability distance, in that case. Unfortunately, this is not very interesting, because X is compact.

Exercise 1.4. Show that this distance is compatible with the topology of X.

1.2. State Observers

1.2.1. Definitions

Definition 1.3. An $U^{r,B}$ state observer of Σ, relative to Ω, is a system $(\Sigma_{Ox}^{r,B,\Omega})$:

$$\frac{dz}{dt} = F\left(z, u^{(0)}, \tilde{u}_r, y^{(0)}\right), \ \eta = \mathcal{H}\left(z, u^{(0)}, \tilde{u}_r, y^{(0)}\right), \tag{70}$$

on the d_Z dimensional manifold Z, where $\Omega \subset X$ is open, η, the output, belongs to X, F, and $\mathcal{H}$ are C^∞, and satisfy the condition:

For all $u \in U^{r,B}$ and for all $x_0 \in \Omega$, such that the corresponding semi-trajectory of Σ, $x(t, x_0)$, is defined for $t \in [0, +\infty[$ and stays in Ω, and for all $z_0 \in Z$, the output $\eta(t, z_0)$ is well defined and,

$$\lim_{t\to+\infty} d(\eta(t, z_0), x(t, x_0)) = 0, \tag{71}$$

where d is any metric compatible with the topology of X.

Again, **this definition makes sense if Ω is relatively compact only.**

Definition 1.4. An exponential $U^{r,B}$ state observer of Σ, relative to Ω, typically denoted by $(\Sigma_{Oxe}^{r,B,\Omega})$, is a one parameter family of state observers for Σ, depending on the real parameter $\alpha > 0$ (on the same manifold Z), which satisfies the following condition (72), strengthening (71):

For any compact $\bar{K} \subset Z$, for any Riemannian distance d on X, there exists $a > 0$, and $k : R+ \to R+$ with polynomial growth, k and a depending on d and $\bar{K}$, such that:

For all $x_0 \in \Omega$, for all $u \in U^{r,B}$, for all $z_0 \in \bar{K}$,

$$Inf[a, d(\eta(t, z_0), x(t, x_0))] \leq k(\alpha)e^{-\alpha t} d(\eta(0, z_0), x_0), \tag{72}$$

as long as $x(t, x_0)$ stays in Ω.

It is important to note that, in the preceding definition, one can replace "there exists $a > 0$" by "for all a, $0 < a \leq a_0$."

Again, if X is not compact, and Ω is not relatively compact, the inequality (72) also does not make sense: All Riemannian metrics are not equivalent, hence, the inequality (72) cannot hold for all Riemannian metrics.

Remark 1.3. There is no hope of having a reasonable theory if we ask that condition (72) in Definition 1.4 be valid for any distance on X (compatible with the topology of X), even if Ω is relatively compact. But for Riemannian distances, everything is fine, because differentiable mappings between Riemannian manifolds are locally Lipschitz.

Remark 1.4. The inequality (72) is more complicated than the inequality (67), because of the **peak phenomenon**, well known to engineers (and to control theoretists). In fact, it happens already in the linear theory. We explain it now.

1.2.2. Peak Phenomenon

Assume that Ω is relatively compact. Let d be a given Riemannian metric, and assume that the following inequality is satisfied, instead of (72):

$$d(\eta(t, z_0), x(t, x_0)) \leq k_d(\alpha)e^{-\alpha t} d(\eta(0, z_0), x_0), \qquad (73)$$

where k_d has polynomial growth.

Inequality (73) cannot hold for all Riemannian metrics on X. This is due to the **peak phenomenon**:

It can happen that there exists a trajectory $x(t, x_0)$ of Σ, with $x_0 \in \Omega$, and a function $\alpha \in R_+^* \to t_\alpha \in R_+$, which tends to zero as α tends to $+\infty$, and such that, if $\eta_\alpha(t, z_0)$ denotes a corresponding trajectory of the observer, $\eta_\alpha(t_\alpha, z_0) \to \infty$ as α tends to $+\infty$.

One can construct a Riemannian metric on X such that for the associated distance function δ on X, $\delta(\eta_\alpha(t_\alpha, z_0), x(t_\alpha, x_0))$ tends to $+\infty$ faster than any power of α.

The peak phenomenon already occurs in the linear theory for the classical Luenberger or Kalman observers. Of course, estimations of x that do not belong to Ω are irrelevant, but this is unimportant. The only important point is that relevant estimations of x are obtained in arbitrary short time, for α large enough.

1.2.3. Consistency of Our Definition of an Exponential State Observer

Exercise 1.5. Show that, on a manifold X, and on any compact set $C \subset X$, the distances induced on C by Riemannian distances are all equivalent.

Lemma 1.1. *If Ω is relatively compact, then Definition 1.4 is independent of the choice of the Riemannian metric d on X.*

Proof. Let K' be the subset of $R^{(r+1)d_u}$ formed by the $(u^{(0)}, \tilde{u}_r)$ that have components bounded by B. Let K'' be the set: $K'' = \cup\{h(\Omega\ , u^{(0)});\ |u_i^{(0)}| \leq B\}$. Let $K''' = \mathcal{H}(\bar{K} \times K' \times K'')$. This set K''' is relatively compact.

Consider Ω' to be open, relatively compact, $cl(\Omega) \subset \Omega'$.

Let δ be the distance associated to another Riemannian metric over X. Then, by the previous exercise, there are λ, $\mu > 0$, such that $\lambda\, d(x, y) \leq \delta(x, y) \leq \mu\, d(x, y)$, for any $(x, y) \in cl(\Omega' \cup K''')$. Set $a_\delta = \delta(\Omega, X\backslash\Omega') = \delta(\partial\Omega, \partial\Omega')$, $a_d = d(\Omega, X\backslash\Omega') = d(\partial\Omega, \partial\Omega')$. One has $a_\delta \leq \mu\ a_d$. Chose Ω' small enough for $a_d \leq a$.

We want to show that, under the conditions of the definition:

$$Inf[a_\delta, \delta(\eta(t, z_0), x(t, x_0))] \leq \bar{k}(\alpha)e^{-\alpha t}\delta(\eta(0, z_0), x_0),$$

for a certain function $\bar{k}$, with polynomial growth.

If $\eta(t, z_0) \notin \Omega'$, then $\delta(\eta(t, z_0), x(t, x_0)) \geq a_\delta$, $d(\eta(t, z_0), x(t, x_0)) \geq a_d$, and

$$Inf[a_\delta, \delta(\eta(t, z_0), x(t, x_0))] = a_\delta \leq \mu\, a_d = \mu Inf[a_d, d(\eta(t, z_0), x(t, x_0))].$$

If $\eta(t, z_0) \in \Omega'$, then, $\delta(\eta(t, z_0), x(t, x_0)) \leq \mu\, d(\eta(t, z_0), x(t, x_0))$, hence,

$$Inf[a_\delta, \delta(\eta(t, z_0), x(t, x_0))] \leq \mu Inf[a_d, d(\eta(t, z_0), x(t, x_0))].$$

Therefore, for any $\eta(t, z_0)$,

$$\begin{aligned} Inf[a_\delta, \delta(\eta(t, z_0), x(t, x_0))] &\leq \mu Inf[a_d, d(\eta(t, z_0), x(t, x_0))] \\ &\leq \mu Inf[a, d(\eta(t, z_0), x(t, x_0))] \leq \mu k(\alpha)e^{-\alpha t}d(\eta(0, z_0), x_0) \\ &\leq \frac{\mu}{\lambda}k(\alpha)e^{-\alpha t}\delta(\eta(0, z_0), x_0). \end{aligned}$$

This shows the result, with $\bar{k} = \frac{\mu}{\lambda}k$. ■

1.2.4. Alternative Definitions of an Exponential State Observer

We give now two other **apparently different**, but more tractable, definitions than Definition 1.4.

Definition 1.5. An exponential $U^{r,B}$ state observer of Σ, relative to Ω, is a one parameter family of state observers for Σ, depending on the real parameter $\alpha > 0$ (on the same manifold Z, independent of α), which satisfies the following condition:

There exists a Riemannian metric d such that relation (73) holds for all $x_0 \in \Omega$, for all $\bar{K}$ compact, for all $z_0 \in \bar{K}$, and for all $u \in U^{r,B}$, as long as $x(t, x_0)$ stays in Ω. The function k_d has polynomial growth and depends on $\bar{K}$.

Definition 1.6. An exponential $U^{r,B}$ state observer of Σ, relative to Ω, is a one parameter family of state observers for Σ, depending on the real parameter $\alpha > 0$ (on the same manifold Z, independent of α), which satisfies the following condition:

There exists a **Riemannian metric** d such that relation (73) holds for all $x_0 \in \Omega$, for all $z_0 \in Z$, and for all $u \in U^{r,B}$, as long as $x(t, x_0)$ stays in Ω. The function k_d has polynomial growth.

Of course, Definition 1.6 is strictly contained in Definition 1.5. On the other hand, if Ω is relatively compact, by Lemma 1.1, Definition 1.5 is itself contained in Definition 1.4. But, in fact, we have the following proposition.

Proposition 1.2. *Definitions 1.4 and 1.5 are equivalent.*

The only motive for introducing the equivalent Definition 1.4 is that it is independent of any special Riemannian metric over X.

Proof. (Proposition 1.2.). Take a Riemannian metric d over X such that X has finite diameter D.

Exercise 1.6. Prove that such a Riemannian metric d does exist.

Take the a from Definition 1.4 applied to the metric d, and a real $\lambda > 1$, with $D < \lambda a$. Consider two trajectories in X, $x(t, x_0)$, $\eta(t, z_0)$ of the system and of the observer system, $x_0 \in \Omega$, $z_0 \in \bar{K}$.

If $Inf[a, d(\eta(t, z_0), x(t, x_0))] = d(\eta(t, z_0), x(t, x_0))$, then

$$\begin{aligned} d(\eta(t, z_0), x(t, x_0)) &\leq k(\alpha)e^{-\alpha t} d(\eta(0, z_0), x_0), \\ &\leq \lambda k(\alpha)e^{-\alpha t} d(\eta(0, z_0), x_0). \end{aligned}$$

If $Inf[a, d(\eta(t, z_0), x(t, x_0))] = a$, then

$$d(\eta(t, z_0), x(t, x_0)) \leq D < \lambda a \leq \lambda k(\alpha)e^{-\alpha t} d(\eta(0, z_0), x_0).$$

Then, in all cases, $d(\eta(t, z_0), x(t, x_0)) \leq \lambda k(\alpha)e^{-\alpha t} d(\eta(0, z_0), x_0)$. This shows that (73) holds with $k_d = \lambda k$. ■

The observers that we will construct will be of two types, as in the classical linear theory: (1) Luenberger type and (2) Kalman's type. For the Luenberger case, the statement of Definition 1.6 is valid, and in the Kalman's case, the statement of Definition 1.5 is valid. This is true both for the classical linear theory and our nonlinear theory.

1.3. Relations between State Observers and Output Observers

1. Assume that $\Sigma_{Oy}^{r,B,\Omega}$ is an **output** observer. If $X = R^n$ and if Ω is relatively compact, then, by Lemma 5.1 and the comment just after, in Chapter 5, there is a continuous function $\varphi_N^{\Omega} : R^{Nd_y} \times R^{Nd_u} \to X$, such that $x = \varphi_N^{\Omega}(y, \tilde{y}_{N-1}, u, \tilde{u}_{N-1})$. This function provides a

continuous single-valued estimation $\hat{x}(t)$ of $x(t, x_0) : \hat{x}(t) = \varphi_N^{\Omega}(\eta(t), u(t), \tilde{u}_{N-1}(t))$. If $|.|$ denotes any norm on $X = R^n$, one has

$$\lim_{t \to +\infty} |x(t) - \hat{x}(t)| = 0. \tag{74}$$

Therefore, in that case, the output observer $\Sigma_{Oy}^{r,B,\Omega}$ allows to construct an $U^{r,B}$ state observer, relative to Ω.

2. The assumptions are the same as in 1 but Σ is **strongly** differentially observable of order N, and (67) holds. Then, the function φ_N^{Ω} can be taken smooth, compactly supported, and if d is any Riemannian distance over X, $d(x(t), \hat{x}(t)) \leq k_d(\alpha)\, e^{-\alpha t}$ for some function k_d with polynomial growth, depending on d, Ω , $\bar{K}$.
3. Assume that $\Sigma_{Ox}^{r',B,\Omega}$, $r' \geq 0$ is a **state** observer. Then, a fortiori, it is an $U^{r,B}$ state observer for some r_0, for all $r \geq r_0 \geq N$. We know that Ω is relatively compact. We can use the mapping Φ_r^{Σ} in order to construct an $U^{r,B}$ output observer as follows: we can replace Φ_r^{Σ} by a smooth (C^{∞}) mapping $\tilde{\Phi}_r^{\Sigma}$, which is constant outside a compact set, and the restriction of which to $\Omega \times V$ coincides with Φ_r^{Σ}. Here, V is the set of $(r-1)$ jets at $t = 0$ of control functions $u(t)$, the $r-1$ first derivatives of which are bounded by B ($V = (U \cap (I_B)^{d_u}) \times (I_B)^{(r-1)d_u}$). Taking the composition $\tilde{\mathcal{H}}$ of this mapping $\tilde{\Phi}_r^{\Sigma}$ with $\mathcal{H}$, as the output mapping of the observer, we get an $U^{r,B}$ output observer relative to Ω.
4. If $\Sigma_{Ox}^{r,B,\Omega}$ is exponential, then: i. $\|\tilde{\Phi}_r^{\Sigma}(\eta(t, z_0), \underline{u}_r(t)) - \tilde{\Phi}_r^{\Sigma}(x(t, x_0), \underline{u}_r(t))\| \leq \lambda d(\eta(t, z_0), x(t, x_0))$ for λ large enough. ii. $\|\tilde{\Phi}_r^{\Sigma}(\eta(t, z_0), \underline{u}_r(t)) - \tilde{\Phi}_r^{\Sigma}(x(t, x_0), \underline{u}_r(t))\| \leq M$ because $\tilde{\Phi}_r^{\Sigma}$ is single-valued outside a compact. Hence, for λ large enough:

$$\|\tilde{\Phi}_r^{\Sigma}(\eta(t, z_0), \underline{u}_r(t)) - \tilde{\Phi}_r^{\Sigma}(x(t, x_0), \underline{u}_r(t))\| \leq \lambda\, Inf\left(d(\eta(t, z_0), x(t, x_0)), \frac{M}{\lambda} \right),$$

$$\|\tilde{\Phi}_r^{\Sigma}(\eta(t, z_0), \underline{u}_r(t)) - \tilde{\Phi}_r^{\Sigma}(x(t, x_0), \underline{u}_r(t))\| \leq \lambda\, k(\alpha) e^{-\alpha t} d(\eta(0, z_0), x_0).$$

Exercise 1.7. ($d_y > d_u$, $r \geq 2n + 1$). If moreover Σ is **strongly differentially observable** of order r, prove that $\tilde{\Phi}_r^{\Sigma}$ can be chosen so that $d(\eta(0, z_0), x_0) \leq \gamma\ \|\tilde{\Phi}_r^{\Sigma}(\eta(0, z_0), \underline{u}_r(0)) - \tilde{\Phi}_r^{\Sigma}(x(0, x_0), \underline{u}_r(0))\|$, for all $z_0 \in \bar{K}$ and all $x_0 \in \Omega$, where γ depends on the compact $\bar{K}$. This shows that the modified observer is an $U^{r,B}$ exponential output observer.

2. The High-Gain Construction

2.1. Discussion about the High-Gain Construction

The **high-gain construction** is a general way to construct either state or output $U^{r,B}$ observers, that are also **exponential**. Before explaining this construction, we want to point out a certain number of facts, concerning the results in this chapter.

1. **Systems with a phase variable representation:** For the systems appearing in Chapters 4 and 5, we obtain a phase variable representation of a certain order N. As we shall see, our **high-gain observers** work for systems in the phase variable representation for C^N controls. Hence, these apply to these general classes of systems.
2. **Systems with a uniform canonical flag** (the single output controlled case of Chapter 3): In that case, as we know (see Exercise 5.4 of Chapter 5), Σ satisfies the phase variable property of some order, either locally or in restriction to arbitrarily large compact sets if Σ is also differentially observable.
   Hence, the construction also applies for sufficiently smooth inputs. **But there is a stronger result: If Σ has the observability canonical form** (20) of Chapter 3, then **our observers work also for arbitrary L^∞ inputs**, i.e., they are $U^{0,B}$ observers.
3. **The high gain construction has mainly two versions**: Referring to the terminology of linear systems theory, the first version is in the **Luenberger style** and the second version is in the **Kalman filter style** (a deterministic version of the Kalman filter).
4. **There are versions of the "high gain observer" that are "continuous–continuous", and others that are "continuous–discrete":** Continuous–continuous means that the observer equation are ordinary differential equations (ODE's), and observations are continuous functions of time. In the continuous–discrete version, which is more realistic, the observer equations are ODE's with jumps, and observations are sampled.

2.2. The Luenberger Style Observer

This section will concern uniformly infinitesimally observable systems. Let us assume that $X = R^n$ and our system Σ, **analytic**, has the observability canonical form (20) **globally**. Recall that this canonical form exists locally as soon as Σ has a uniform canonical flag.

Let us denote by $\underline{x}_i$ the vector $(x^0, \ldots, x^i)$. The following two additional assumptions will be crucial.

(A_1) Each of the maps f_i, $i = 0, \dots, n-1$, is globally Lipschitz w.r.t. $\underline{x}_i$, uniformly with respect to u and x^{i+1},

(A_2) there exists two real α, β, $0 < \alpha < \beta$, such that

$$\begin{aligned} \alpha &\leq \left| \frac{\partial h}{\partial x^0} \right| \leq \beta, \\ \alpha &\leq \left| \frac{\partial f_i}{\partial x^{i+1}} \right| \leq \beta, 0 \leq i \leq n-2. \end{aligned} \tag{75}$$

In fact, these assumptions can be automatically satisfied, as soon as one is interested only in the trajectories that stay in a given compact convex set $\Gamma \subset X = R^n$, as shown in the following exercise.

Exercise 2.1. Assume that $X = R^n$, $\Gamma \subset X$ is the closure of an open, relatively compact, **convex** subset of X, and Σ has the normal form (20), (it is sufficient that it has this normal form only on Γ).

Show that for all $B > 0$, Σ can be extended smoothly (C^∞) outside of $\Gamma \times V_B$, so that the assumptions A_1, A_2, are satisfied (globally) on $X \times U$. (Here, $V_B = \{u \in U; |u| \leq B\}$.)

In order to prove the main result of this section, we will need the following technical lemma:

Lemma 2.1. *Consider time-dependant real matrices $A(t)$ and $C(t)$:*

$$A(t) = \begin{pmatrix} 0, \varphi_2(t), \dots\dots, 0 \\ 0, 0, \varphi_3(t), \dots\dots, 0 \\ \cdot \\ \cdot \\ 0, 0, \dots\dots, 0, \varphi_n(t) \\ 0, 0, \dots\dots\dots, 0, 0 \end{pmatrix}, \quad C(t) = (\varphi_1(t), 0, \dots, 0).$$

$A(t)$ is $n \times n$, and $C(t)$ is $1 \times n$. Assume that there are two real constant α, β, such that

$$0 < \alpha < \beta, \alpha < \varphi_i(t) < \beta, 1 \leq i \leq n.$$

Then, there is a real $\lambda > 0$, a vector $\bar{K} \in R^n$, and a symmetric positive definite $n \times n$ matrix S, λ and S depending on α, β only, such that

$$(A(t) - \bar{K}C(t))'S + S(A(t) - \bar{K}C(t)) \leq -\lambda\, Id.$$

Here, $(A(t) - \bar{K}C(t))'$ means transpose of $(A(t) - \bar{K}C(t))$ and $\leq$ is the (partial) ordering of symmetric matrices defined by the cone of symmetric positive semidefinite matrices.

Proof. (Personal communication from W. Dayawansa). The proof is an induction on the dimension n. For $n = 1$, we consider the quadratic form $S(x, x) = \frac{x^2}{2}$. Then, $S(x, (A - \bar{K}C)x) = -\bar{K}\varphi_1(t)\frac{x^2}{2}$, and, if an arbitrary $\lambda > 0$ is given, any real $\bar{K}$ sufficiently large does the job.

Step n: We are looking for $\bar{K} = (k, K)$, $k \in R$, $K \in R^{n-1}$, and we have to consider:

$$\left[A(t) - \begin{pmatrix} k \\ K \end{pmatrix} C(t)\right]x = \begin{bmatrix} -k\varphi_1, C_1(t) \\ -K\varphi_1, A_1(t) \end{bmatrix} \begin{pmatrix} x_1 \\ x_2 \end{pmatrix}, x_1 \in R, x_2 \in R^{n-1}, \tag{76}$$

with $C_1(t) = (\varphi_2(t), 0, \ldots, 0)$, and

$$A_1(t) = \begin{pmatrix} 0, \varphi_3(t), \ldots\ldots, 0 \\ 0, 0, \varphi_4(t), \ldots, 0 \\ \vdots \\ 0, 0, \ldots, 0, \varphi_n(t) \\ 0, 0, \ldots\ldots, 0, 0 \end{pmatrix}.$$

First, we make the following linear coordinate change:

$$z_1 = x_1, z_2 = x_2 + \Omega x_1,$$

where $\Omega \in R^{n-1}$.

The square matrix appearing in the right-hand side of (76) becomes

$$\begin{pmatrix} -k\varphi_1 - C_1\Omega & C_1 \\ -(K + \Omega k)\varphi_1 - (A_1 + \Omega C_1)\Omega & (A_1 + \Omega C_1) \end{pmatrix} = B(t). \tag{77}$$

By the induction hypothesis relative to α and β, there is a $\lambda > 0$, an Ω, and a quadratic Lyapunov function $z_2' S_{n-1} z_2$ such that

$$(A_1 + \Omega C_1)' S_{n-1} + S_{n-1}(A_1 + \Omega C_1) \leq -\lambda\, Id.$$

We will look for S_n of the form $\begin{pmatrix} \frac{1}{2} & 0 \\ 0 & S_{n-1} \end{pmatrix}$. Setting $V(Z, Y) = Z' S_n Y$, we get

$$\begin{aligned} 2V(Z, B(t)Z) &= (-k\varphi_1 - C_1\Omega)z_1^2 + C_1 z_2 z_1 + 2z_2' S_{n-1}(A_1 + \Omega C_1)z_2 \\ &\quad + 2z_2' S_{n-1}[-(K + \Omega k)\varphi_1 - (A_1 + \Omega C_1)\Omega]z_1, \\ 2V(Z, B(t)Z) &\leq (-k\varphi_1 - C_1\Omega)z_1^2 - \lambda\|z_2\|^2 \\ &\quad + (\varphi_2 + 2\|S_{n-1}\|\|(K + \Omega k)\varphi_1 + (A_1 + \Omega C_1)\Omega\|)|z_1|\|z_2\|. \end{aligned}$$

We can choose $K = -\Omega k$, and k is to be determined. For any $\varepsilon > 0$,

$$|z_1|\ \|z_2\| = |z_1/\varepsilon|\ \|\varepsilon z_2\| \leq \frac{1}{2}\left(\varepsilon^2\|z_2\|^2 + \frac{1}{\varepsilon^2}|z_1|^2\right).$$

Hence,

$$2V(Z, B(t)Z) \leq \left((-k\varphi_1 - C_1\Omega) + \frac{\delta(t)}{2\varepsilon^2}\right)|z_1|^2 + \left(-\lambda + \delta(t)\frac{\varepsilon^2}{2}\right)\|z_2\|^2,$$

where $\delta(t) = \varphi_2 + 2\|S_{n-1}\|\ \|(A_1 + \Omega C_1)\Omega\|$.

Because $\delta(t)$ is bounded from above, ε can be chosen small enough in order that $(-\lambda + \delta(t)\frac{\varepsilon^2}{2}) < -\frac{\lambda}{2}$. Because φ_1 is bounded from below, k can be chosen large enough in order that $(-k\varphi_1 - C_1\Omega) + \frac{\delta(t)}{2\varepsilon^2} < -\frac{\lambda}{2}$. Hence, $2V(Z, B(t)Z) \leq -\frac{\lambda}{2}\|Z\|^2$. Setting $\theta = \begin{pmatrix} 1 & 0 \\ \Omega & Id \end{pmatrix}$, we have $Z = \theta x$, $2x'\theta' S_n B(t)\theta x \leq -\frac{\lambda}{2}x'\theta'\theta x$. Hence, setting $\tilde{S}_n = \theta' S_n \theta$, we have

$$2x'\tilde{S}_n\left[A(t) - \begin{pmatrix} k \\ K \end{pmatrix} C(t)\right]x \leq -\frac{\lambda}{2}x'\theta'\theta x,$$

and we obtain

$$2x'\tilde{S}_n\left[A(t) - \begin{pmatrix} k \\ K \end{pmatrix} C(t)\right]x \leq -\frac{\lambda}{2}\gamma\|x\|^2,$$

for some $\gamma > 0$, which ends the proof. ∎

Now, let us define the dynamics of our **observer system** Σ_O as follows:

$$\frac{d\hat{x}}{dt} = f(\hat{x}, u) - K_\theta(h(\hat{x}, u) - y), \tag{78}$$

where $K_\theta = \Delta_\theta K$, $\Delta_\theta = diag(\theta, \theta^2, \ldots, \theta^n)$ for $\theta > 1$, and K (together with S and λ) comes from Lemma 2.1, relative to α, β, in the assumption A_2, (75).

Theorem 2.2. *For any $a > 0$, there is a $\theta > 1$ (large enough) such that*

$$\forall (x_0, \hat{x}_0) \in X \times X,\ \|\hat{x}(t) - x(t)\| \leq k(a)e^{-at}\|\hat{x}_0 - x_0\|, \tag{79}$$

for some polynomial k, of degree n, where $\hat{x}(t)$ and $x(t)$ denote the solutions at time t of the observer system Σ_O and the system Σ, with respective initial conditions $\hat{x}_0$, x_0.

Corollary 2.3. *For all $B > 0$, for any relatively compact $\Omega \subset X$, the system Σ_O given by Formula (78) is an $U^{0,B}$ exponential state observer for Σ, relative to Ω.*

Corollary 2.4. *Let Ω be an open relatively compact convex subset of $X = R^n$. Assume that the restriction $\Sigma_{|cl(\Omega)}$ is* **globally** *in the observability canonical form (20). Then, for all $B > 0$, there is a $U^{0,B}$ exponential state observer for Σ, relative to Ω.*

Observation: in both corollaries, the polynomial $k(a)$ is independent of the compact set $\bar{K}$ in the definition of the observer.

Corollary 2.3 is an immediate consequence of Theorem 2.2 and of the definition of a state observer. Corollary 2.4 is a consequence of Exercise 2.1 and of Theorem 2.2.

Proof of Theorem 2.2. Our system Σ is written as

$$(\Sigma)\quad \begin{pmatrix} \dot{x}_1 \\ \cdot \\ \cdot \\ \dot{x}_n \end{pmatrix} = \begin{pmatrix} \varphi_1(x_1, x_2, u), \\ \cdot \\ \cdot \\ \varphi_n(x_1, \dots, x_n, u), \end{pmatrix} = f(x, u).$$
$$y = \varphi_0(x_1, u).$$

The equation of the observer is

$$\frac{d\hat{x}}{dt} = f(\hat{x}, u) - K_\theta(\varphi_0(\hat{x}, u) - y).$$

Setting $\varepsilon = \hat{x} - x$, we get

$$\dot{\varepsilon} = f(\hat{x}, u) - f(x, u) - K_\theta(\varphi_0(\hat{x}, u) - y). \tag{80}$$

We consider trajectories $x(t)$, $\hat{x}(t)$, $\varepsilon(t)$ of (Σ), (Σ_O), (80) respectively, corresponding to the control function $u(t)$. We denote $\underline{x}_i$ for $(x_1, \dots, x_i)$. One has

$$\begin{aligned}
\varphi_i(\hat{x}, u) - \varphi_i(x, u) &= \varphi_i(\underline{\hat{x}}_i, \hat{x}_{i+1}, u) - \varphi_i(\underline{x}_i, x_{i+1}, u) \\
&= \varphi_i(\underline{\hat{x}}_i(t), \hat{x}_{i+1}(t), u(t)) - \varphi_i(\underline{x}_i(t), \hat{x}_{i+1}(t), u(t)) \\
&\quad + \varphi_i(\underline{x}_i(t), \hat{x}_{i+1}(t), u(t)) - \varphi_i(\underline{x}_i(t), x_{i+1}(t), u(t)) \\
&= \varphi_i(\underline{\hat{x}}_i(t), \hat{x}_{i+1}(t), u(t)) - \varphi_i(\underline{x}_i(t), \hat{x}_{i+1}(t), u(t)) \\
&\quad + \frac{\partial \varphi_i}{\partial x_{i+1}}(\underline{x}_i(t), \delta_i(t), u(t))\varepsilon_{i+1}(t),
\end{aligned}$$

for some $\delta_i(t)$.

We set

$$g_{i+1}(t) = \frac{\partial \varphi_i}{\partial x_{i+1}}(\underline{x}_i(t), \delta_i(t), u(t)),$$

to obtain

$$\dot{\varepsilon}_i = \varphi_i(\underline{\hat{x}}_i(t), \hat{x}_{i+1}(t), u(t)) - \varphi_i(\underline{x}_i(t), \hat{x}_{i+1}(t), u(t)) + g_{i+1}(t)\varepsilon_{i+1}(t)$$
$$- (K_\theta)_i \, g_1(t)\varepsilon_1,$$

where $g_1(t) = \frac{\partial \varphi_0}{\partial x_1}(\delta_0(t), u(t))$. Hence,

$$\dot{\varepsilon} = (A(t) - K_\theta C(t))\varepsilon + \bar{F},$$

with $C(t) = (g_1(t), 0, \ldots, 0)$,

$$A(t) = \begin{pmatrix} 0, g_2(t), 0, \ldots, , 0 \\ 0, 0, g_3(t), 0, \ldots, 0 \\ \cdot \\ 0, \ldots\ldots\ldots\ldots, g_n(t) \\ 0, 0, \ldots\ldots\ldots\ldots, 0 \end{pmatrix},$$

and

$$\bar{F} = \begin{pmatrix} \cdot \\ \cdot \\ \varphi_i(\underline{\hat{x}}_i(t), \hat{x}_{i+1}(t), u(t)) - \varphi_i(\underline{x}_i(t), \hat{x}_{i+1}(t), u(t)) \\ \cdot \\ \cdot \end{pmatrix}.$$

The assumption A_1 states that the φ_i are Lipschitz with respect to $\underline{x}_i$. We set

$$x^0 = (\Delta_\theta)^{-1}x, \ \hat{x}^0 = (\Delta_\theta)^{-1}\hat{x}, \ \varepsilon^0 = (\Delta_\theta)^{-1}\varepsilon.$$

We have

$$\dot{\varepsilon}^0 = (\Delta_\theta)^{-1}\dot{\varepsilon} = (\Delta_\theta)^{-1}(A(t) - K_\theta C(t))\Delta_\theta \varepsilon^0 + (\Delta_\theta)^{-1}\bar{F}.$$

Otherwise,

$$(\Delta_\theta)^{-1}\bar{F} = (\Delta_\theta)^{-1} \begin{pmatrix} \cdot \\ \cdot \\ \varphi_i\left(\Delta_\theta \underline{\hat{x}}_i^0(t), \hat{x}_{i+1}(t), u(t)\right) - \varphi_i\left(\Delta_\theta \underline{x}_i^0(t), \hat{x}_{i+1}(t), u(t)\right) \\ \cdot \\ \cdot \end{pmatrix}.$$

The main fact is that $\|(\Delta_\theta)^{-1}\bar{F}\| \leq L\|\varepsilon^0\|$, where $\frac{L}{\sqrt{n}}$ is the Lipschitz constant of the φ_i's with respect to $\underline{x}_i$ (an easy verification left to the reader.)

Hence, using Lemma 2.1,

$$\frac{1}{2}\frac{d}{dt}(\varepsilon^{0\prime}S\varepsilon^0) = \varepsilon^{0\prime}S\dot{\varepsilon}^0 = \varepsilon^{0\prime}S(\Delta_\theta)^{-1}(A(t) - K_\theta C(t))\Delta_\theta\varepsilon^0 + \varepsilon^{0\prime}S(\Delta_\theta)^{-1}\bar{F}$$

$$\leq \theta\varepsilon^{0\prime}S(A(t) - KC(t))\varepsilon^0 + L\|S\|\ \|\varepsilon^0\|^2$$

$$\leq \left(-\theta\frac{\lambda}{2} + L\|S\|\right)\|\varepsilon^0\|^2.$$

Hence, for an arbitrary $\gamma > 0$, if θ is sufficiently large,

$$\frac{d}{dt}(\varepsilon^{0\prime}S\varepsilon^0) \leq -\gamma\varepsilon^{0\prime}S\varepsilon^0,$$

which gives $\varepsilon^{0\prime}S\varepsilon^0 \leq \varepsilon^{0\prime}S\varepsilon^0(0)e^{-\gamma t}$. The result follows immediately. ■

2.3. The Case of a Phase-Variable Representation

Here, d_y is arbitrary. Let us assume that we have a system in the phase-variable representation, $y^{(N)} = \varphi(y^{(0)}, \tilde{y}_{N-1}, u^{(0)}, \tilde{u}_N)$, then the previous construction can be adapted to obtain an exponential $U^{N,B}$ output observer, at the cost of an additional assumption:

(A_3) φ is compactly supported w.r.t. $(y^{(0)}, \tilde{y}_{N-1})$. (Remember that this means that for any K', a compact subset of $U \times R^{Nd_u}$, the restriction of φ to $R^{Nd_y} \times K'$ is compactly supported.)

As we know, this assumption is satisfied in many situations, for instance:

- In Chapter 5, in the situation of Theorem 5.2,
- In particular, in the case where $d_y > d_u$, we have generically a phase-variable representation. If we restrict Σ to a compact subset $\bar{\Omega}$ of X, Theorem 5.2 gives us a phase-variable representation for Σ, with the additional property (A_3).

Let us denote by A the (Nd_y, d_y) block-antishift matrix:

$$A = \begin{pmatrix} 0, Id_{d_y}, 0, \ldots\ldots\ldots, 0 \\ 0, 0, Id_{d_y}, 0, \ldots\ldots, 0 \\ 0, \\ 0, 0, \ldots\ldots\ldots, 0, Id_{d_y} \\ 0, 0, \ldots\ldots\ldots\ldots, 0, 0 \end{pmatrix}.$$

Here, $\hat{z}$ denotes a typical element of R^{Nd_y}. Then $\frac{d\hat{z}}{dt} = A\hat{z}$ is the linear ODE on R^{Nd_y} corresponding to the vector field

$$A(\hat{z}) = \sum_{\substack{i=1,\ldots,N-1\\ j=1,d_y}} \left\{ \hat{z}_{id_y+j} \frac{\partial}{\partial \hat{z}_{(i-1)d_y+j}} \right\}.$$

Also, $b(\hat{z}, u^{(0)}, \tilde{u}_N)$ denotes a vector field on R^{Nd_y}, depending on $u^{(0)}, \tilde{u}_N$, defined as follows:

$$b\big(\hat{z}, u^{(0)}, \tilde{u}_N\big) = \sum_{j=1}^{d_y} \varphi_j\big(\hat{z}, u^{(0)}, \tilde{u}_N\big) \frac{\partial}{\partial \hat{z}_{(N-1)d_y+j}}.$$

$C : R^{Nd_y} \to R^{d_y}$ is the linear mapping with matrix:

$$C = (Id_{d_y}, 0, \ldots, 0),$$

i.e., $C(\hat{z})$ denotes the vector of R^{d_y} the components of which are the d_y first components of $\hat{z}$.

Definition 2.1. A square matrix A is called stable if all its eigenvalues have strictly negative real parts.

Consider the system Σ_O, on R^{Nd_y}, with output $\eta \in R^{Nd_y}$,

$$(\Sigma_O)\ \frac{d\hat{z}}{dt} = A(\hat{z}) + b\big(\hat{z}, u^{(0)}, \tilde{u}_N\big) - K_\theta(C(\hat{z}) - y(t)), \eta = \hat{z}, \quad (81)$$

where $\theta > 1$ is a given real, $K_\theta = \Delta_\theta K$, Δ_θ is the block diagonal matrix,

$$\Delta_\theta = Block - diag(\theta Id_{d_y}, \theta^2 Id_{d_y}, \ldots, \theta^N Id_{d_y}),$$

and K is such that the matrix $A - KC$ is a stable matrix.

Exercise 2.2. Prove that such a K does exist.

Theorem 2.5. *Σ_O is an exponential $U^{N,B}$ output observer relative to an arbitrarily large relatively compact $\Omega \subset X$ (if the assumption A_3 is satisfied; if not, one has first to modify the mapping φ outside $S\Phi_N^\Sigma(\Omega \times R^{Nd_u})$ in order that A_3 be satisfied in which case, the observer system depends on Ω).*

Proof. This proof is just an adaptation of the proof of Theorem 2.2. It gives a more simple and explicit construction of the parameters K_θ of the observer. Let us do it in detail again. First, we find a symmetric positive definite matrix S by solving the Lyapunov equation:

$$(A - KC)'S + S(A - KC) = -Id. \quad (82)$$

■

Exercise 2.3. Show that this equation has a single symmetric positive definite solution S.

We know that, because Σ has the phase-variable property of order N, the trajectories of Σ (corresponding to sufficiently differentiable controls) are mapped by Φ_N^Σ into the trajectories of the system

$$(\Sigma')\ \dot{z} = A(z) + b\big(z, u^{(0)}(t), \tilde{u}_N(t)\big). \tag{83}$$

If the phase-variable property holds in restriction to a relatively compact subset $\Omega \subset X$ only, this last property holds for the (pieces of) semitrajectories that stay in Ω only.

We consider any trajectory of Σ, corresponding to the initial condition x_0 and the control $u(t)$, $t \geq 0$, such that $|\frac{d^i u}{dt^i}| \leq B$, $0 \leq i \leq N$. Then $z(0) = z_0 = \Phi_N^\Sigma(x_0, u^{(0)}(0), \tilde{u}_{N-1}(0))$. The corresponding trajectory of Σ' is just $z(t) = (y^{(0)}(t), \tilde{y}_{N-1}(t))$, the successive time derivatives of the output of Σ (as long as $x(t) \in \Omega$). We fix an arbitrary initial condition $\hat{z}(0) = \hat{z}_0$. We set $\varepsilon(t) = \hat{z}(t) - z(t)$. By construction, ε satisfies

$$\frac{d\varepsilon}{dt} = (A - K_\theta C)\varepsilon(t) + b\big(\hat{z}(t), u^{(0)}(t), \tilde{u}_N(t)\big) - b\big(z(t), u^{(0)}(t), \tilde{u}_N(t)\big). \tag{84}$$

For the sake of simplicity of the notations, we will omit the variables $u^{(0)}(t), \tilde{u}_N(t)$ from now on and rewrite this equation as

$$\frac{d\varepsilon}{dt} = (A - K_\theta C)\varepsilon(t) + b(\hat{z}(t)) - b(z(t)). \tag{85}$$

Set $\varepsilon = \Delta_\theta \varepsilon^\circ$, $z = \Delta_\theta z^\circ$, $\hat{z} = \Delta_\theta \hat{z}^\circ$. We obtain

$$\frac{d\varepsilon^\circ}{dt} = \theta(A - KC)\varepsilon^\circ(t) + \frac{1}{\theta^N}(b(\Delta_\theta \hat{z}^\circ(t)) - b(\Delta_\theta z^\circ(t))).$$

The map φ is C^∞, compactly supported in $(y^{(0)}, \tilde{y}_{N-1})$. Hence, it has a Lipschitz constant L_B with respect to $(y^{(0)}, \tilde{y}_{N-1})$, depending on the bound B on the control and its first derivatives. As before, the Lipschitz constant of $\frac{1}{\theta^N} b(\Delta_\theta z)$ is also L_B if $\theta > 1$.

One has

$$\begin{aligned}
\frac{1}{2}\frac{d}{dt}\varepsilon^{\circ\prime} S \varepsilon^\circ &= -\frac{\theta}{2}\varepsilon^{\circ\prime}\varepsilon^\circ + \varepsilon^{\circ\prime} S \frac{1}{\theta^N}(b(\Delta_\theta \hat{z}^\circ(t)) - b(\Delta_\theta z^\circ(t))) \\
&\leq -\frac{\theta}{2}\|\varepsilon^\circ\|^2 + \|S\| L_B \|\varepsilon^\circ\|^2 \\
&\leq -\left(\frac{\theta}{2} - \|S\| L_B\right)\|\varepsilon^\circ\|^2.
\end{aligned}$$

Hence, for $\theta > 2\ L_B\ \|S\|$, then $\frac{1}{2}\frac{d}{dt}\varepsilon^{\circ\prime}S\varepsilon^{\circ} \leq -(\frac{\theta}{2\|S\|} - L_B)\varepsilon^{\circ\prime}S\varepsilon^{\circ}$, and

$$\varepsilon^{\circ\prime}S\varepsilon^{\circ} \leq e^{-2(\frac{\theta}{2\|S\|}-L_B)t}\varepsilon^{\circ\prime}(0)S\varepsilon^{\circ}(0),$$
$$\|\varepsilon^{\circ}\|^2 \leq e^{-2\alpha(\frac{\theta}{2\|S\|}-L_B)t}\|S^{-1}\|\ \|S\|\ \|\varepsilon^{\circ}(0)\|^2.$$

For $\theta > \sup(1, 2\ L_B\ \|S\|)$, we have

$$\|\varepsilon\|^2 = \varepsilon^{\circ\prime}(\Delta_\theta)^2\varepsilon^{\circ} \leq \theta^{2N}\|\varepsilon^{\circ}\|^2$$
$$\leq e^{-2(\frac{\theta}{2\|S\|}-L_B)t}\|S^{-1}\|\ \|S\|\theta^{2N}\ \varepsilon'(0)(\Delta_\theta)^{-2}\varepsilon(0),$$
$$\|\varepsilon\|^2 \leq e^{-2(\frac{\theta}{2\|S\|}-L_B)t}\|S^{-1}\|\ \|S\|\theta^{2(N-1)}\ \|\varepsilon(0)\|^2. \quad \blacksquare$$

Observation: again, the function k in the definition of the exponential output observer does not depend on the compact $\bar{K}$ (in which the observer is initialized). This will not be the case any more in the next paragraph.

2.4. The Extended Kalman Filter Style Construction

2.4.1. Introduction and Main Result

No prerequisites about the Kalman filter are required to understand this section. We remain in the deterministic setting, and we present complete proofs of all results. Let us just add a few **comments** about the **extended Kalman filter**. The "extended Kalman filter" (for simplicity denoted by EKF) applies the linear time-dependent version of the Kalman filter to the linearized system along the **estimate of the trajectory**. If it was along the real trajectory, then the procedure would be perfectly well defined. However, it has to be along the estimate of the trajectory, because the real trajectory is unknown. The purpose of the filter is precisely to estimate it. Therefore, it is an easy exercise to check that the equations of the extended Kalman filter are not intrinsic. They depend on the coordinate system. These equations were introduced by engineers, and they perform very well in practice, because they take the noise into account.

Our point of view in this section is as follows:

1. We use special coordinates, for instance, the special coordinates of the uniform observability canonical form (23) in the single output **control affine** case, or the coordinates of a phase-variable representation in other cases. These coordinates are essentially uniquely defined, hence, **the extended Kalman filter written in these coordinates becomes a well-defined object.**

2. it is possible to adapt the high gain construction shown in the previous section in order that the equations of the EKF in the special coordinates give the same results as in the previous section (arbitrary exponential convergence of the estimation error).

The main difference with the Luenberger style version, is that the **correction term** "K_θ" is not constant: It is computed as a function of the information appearing at the current time t. We have observed in the applications that the EKF performs very well in practice, probably for this reason.

Let us present this construction in the single-output, **control-affine** case: this last requirement seems essential. We make the following assumption:

a1) Σ is globally in the normal form (23).

This is true in several situations, for example if Σ is observable and if we make one of the following assumptions (a2), (a3).

a2) $\Phi = (h, L_f h, \ldots, L_f^{n-1} h)$ is a global diffeomorphism,

or, weaker,

a3) in restriction to Γ, the closure of an open relatively compact subset of X, Φ is a diffeomorphism.

Remember that the observability assumption implies that Φ should be almost everywhere a local diffeomorphism from X into R^n, and that, in the coordinates defined by Φ, the system Σ has to be in the normal form (23). (See Theorem 4.1 of Chapter 3.) In the case of the assumption (a3), all the functions φ and g_i in the normal form (23) can be extended to all of R^n so that they are smooth and compactly supported w.r.t. all their arguments, and g_i depends on $(x_1, \ldots, x_i)$ only.

Let us recall the normal form (23):

$$\dot{x} = \begin{pmatrix} \dot{x}_1 \\ \dot{x}_2 \\ \cdot \\ \cdot \\ \cdot \\ \dot{x}_{n-1} \\ \dot{x}_n \end{pmatrix} = \begin{pmatrix} x_2 \\ x_3 \\ \cdot \\ \cdot \\ \cdot \\ x_n \\ \varphi(x) \end{pmatrix} + u \begin{pmatrix} g_1(x_1) \\ g_2(x_1, x_2) \\ \cdot \\ \cdot \\ \cdot \\ g_{n-1}(x_1, \ldots, x_{n-1}) \\ g_n(x) \end{pmatrix}$$

$$y = x_1$$

Let us assume moreover that:

a4) φ and g_i are globally Lipschitz. In the case of (a3), this will be automatically true, by what we just said.

Denote again by A the antishift matrix:

$$A = \begin{pmatrix} 0, 1, 0, \dots, 0 \\ \dots \\ \dots \\ 0, \dots\dots, 0, 1 \\ 0, 0, \dots, , 0, 0 \end{pmatrix},$$

and let C denote the linear form over R^n with matrix $C = (1, 0, \dots\dots, 0)$. Let us rewrite the normal form (23) in matrix notations as

$$\dot{x} = Ax + b(x, u), \ \ y = Cx, \tag{86}$$

where b_i, the ith component of b depends only on $\underline{x}_i = (x_1, \dots, x_i)$ and u.

Let Q be a given symmetric positive definite $n \times n$ matrix, and r, θ be positive real numbers, $\Delta_\theta = diag(1, \frac{1}{\theta}, \dots, (\frac{1}{\theta})^{n-1})$. Let $b^*(x, u)$ denote the Jacobian matrix of $b(x, u)$ w.r.t. x. Set $Q_\theta = \theta^2(\Delta_\theta)^{-1} Q (\Delta_\theta)^{-1}$. The equations,

$$(i)\ \frac{dz}{dt} = Az + b(z, u) - S(t)^{-1} C' r^{-1}(Cz - y(t)), \tag{87}$$

$$(ii)\ \frac{dS}{dt} = -(A + b^*(z, u))' S - S(A + b^*(z, u)) + C' r^{-1} C - S Q_\theta S,$$

$$\eta = z, \tag{88}$$

define what is called the **extended Kalman filter** for our system (86). (Q_θ and r are analogous to the covariance matrices of the state noise and the output noise in the stochastic context.)

We will show the following:

Theorem 2.6. *Under the assumptions (a1) and (a4), for $\theta > 1$ and for all $T > 0$, the extended Kalman filter (87) satisfies, for $t \geq \frac{T}{\theta}$:*

$$\|z(t) - x(t)\| \leq \theta^{n-1} k(T) \left\| z\left(\frac{T}{\theta}\right) - x\left(\frac{T}{\theta}\right) \right\| e^{-(\theta\omega(T) - \mu(T))(t - \frac{T}{\theta})}, \tag{89}$$

for some positive continuous functions $k(T)$, $\omega(T)$, $\mu(T)$.

Corollary 2.7. *Under the assumptions (a1) and (a4), for any open relatively compact $\Omega \subset X = R^n$ and for any $B > 0$, the extended Kalman filter is an exponential $U^{0,B}$ state observer, relative to Ω.*

In fact, this theorem and this corollary generalize to the case of a multi-output system, having a phase-variable representation. Let us assume that Σ

has the phase-variable representation $y^{(N)} = \varphi(y, \tilde{y}_{N-1}, u, \tilde{u}_N)$, and that φ is compactly supported w.r.t. $(y, \tilde{y}_{N-1})$, as in Section 2.3.

We consider systems on R^{Np}, of the general form

$$\frac{dx}{dt} = A_{N,p}x + b(x, u), \tag{90}$$

where p is an integer, $A_{N,p}$ is the $(Np,\ p)$-antishift matrix:

$$A_{N,p}= \begin{pmatrix} 0, Id_p, 0, \ldots, 0 \\ \cdot \\ \cdot \\ 0, \ldots\ldots, 0, Id_p \\ 0, \ldots\ldots\ldots, 0 \end{pmatrix},$$

and where $\frac{\partial b_i}{\partial x_j} \equiv 0$ if, for some integer k

$$kp < i \leq (k+1)p, \quad \text{and} \quad j > (k+1)p,$$

and all the functions $b_i(x, u)$ are compactly supported, w.r.t. their x arguments.

Clearly, this form **includes not only the normal form** (23), but also the systems with p outputs, which are in the **phase variable representation** (in this case, u **in (90) denotes not only the control, but its** N **first derivatives**).

Let us consider the same extended Kalman filter equations,

$$\begin{aligned} (i)\ \frac{dz}{dt} &= A_{N,p}z + b(z, u) - S(t)^{-1}C'r^{-1}(Cz - y(t)), \\ (ii)\ \frac{dS}{dt} &= -(A_{N,p} + b^*(z, u))'S - S(A_{N,p} + b^*(z, u)) \\ &\quad + C'r^{-1}C - SQ_\theta S, \\ \eta &= z, \end{aligned} \tag{91}$$

where $C = (Id_p, 0, \ldots, 0)$, $Q_\theta = \theta^2\Delta^{-1}Q\Delta^{-1}$, $\Delta = BlockDiag(Id_p, \frac{1}{\theta}Id_p, \ldots, (\frac{1}{\theta})^{N-1}Id_p)$. The expression $b^*(z, u)$ is again the Jacobian matrix of $b(z, u)$ w.r.t. z.

As in the case of our **Luenberger type** observers, we have:

Theorem 2.8. *Theorem 2.6 holds also for systems of the form (90).*

Corollary 2.9. *If a system Σ has a phase-variable representation of order N, then for all open relatively compact subsets $\Omega \subset X$, for all $B > 0$, the system (91) is an exponential $U^{N,B}$ output observer for Σ, relative to Ω.*

Corollary 2.9 is just a restatement of Corollary 2.7 in this new context.

2.4.2. Preparation for the Proof of Theorems 2.6 and 2.8

To prove the theorems, we need a series of lemmas. The first one is a continuity result for control affine systems, very important in itself.

Let Σ be a control affine system (without outputs):

$$(\Sigma) \quad \begin{cases} \dot{x} = f(x) + \Sigma_{i=1}^{p} g_i(x) u_i, \\ x(0) = x_0. \end{cases}$$

Let $P_\Sigma : Dom(P_\Sigma) \subset L^\infty([0,T], R^p) \to C^0([0,T], X)$, be the "input $\to$ state" mapping of Σ, i.e., the mapping that to $u(.)$ associates the state trajectory $t \in [0,T] \to x(t)$.

Let us endow $L^\infty([0,T], R^p)$ with the weak-* topology, and $C^0([0,T], X)$ with the topology of uniform convergence.

Lemma 2.10. *1) The domain of* P_Σ *is open in* $L^\infty([0,T], R^p)$, 2) P_Σ *is continuous on bounded sets.*

Proof. See Appendix 3.1 in this chapter. ■

Let now Σ be a **bilinear** system on R^{Np}, of the form

$$\dot{x} = A_{N,p} x + \sum_{\substack{k,l=0 \\ l \le k}}^{N-1} \sum_{i,j=1}^{p} \{u_{kp+i,lp+j}\, e_{kp+i,lp+j}\, x\}, \tag{92}$$

$$y = Cx,$$

where as above, $A_{N,p}$ is the (Np, p)-antishift matrix; $C : R^{Np} \to R^p$, $C = (Id_p, 0, \ldots, 0)$ is the projection on the p first components; $e_{i,j}$ is the matrix $e_{ij}(v_k) = \delta_{jk} v_i$, where $\{v_j\}$ is the canonical basis of R^{Np}.

The term after $\sum\sum$ in (92) is a lower p-block triangular matrix. For $u(.) = (u_{ij}(.))$, a measurable bounded control function for Σ, defined on $[0,T]$, we denote by $\Psi_u(t,s)$ the associated resolvent matrix. ($\Psi_u(t,s) \in Gl_+(Np, R)$, $s, t \in [0,T]$). Let us define the Gramm observability matrix of Σ by

$$G_u = \int_0^T \Psi_u(\tau, T)' C' C \Psi_u(\tau, T)\, d\tau.$$

The matrix G_u is a symmetric positive semi-definite matrix.

It is easy to check that the bilinear system (92) is observable for all $u(.)$, measurable, bounded.

Lemma 2.11. *If a bound* B *is given on the controls* $u_{i,j}$, *then there exist positive numbers* $0 < \alpha < \beta$, *depending on* B, T *only, such that* $\alpha\, Id_{Np} \le G_u \le \beta\, Id_{Np}$.

Exercise 2.4.

1. Using Lemma 2.10, prove that the map $u(.) \to G_u$ is continuous,
2. Using precompactness of the weak-* topology, and the observability of (92), prove Lemma 2.11.

All the other lemmas in this section deal with properties of solutions of Riccati matrix equations.

Let us set $A(t) = A_{N,p} + \sum_{\substack{k,l=0 \\ l \leq k}}^{N-1} \sum_{i,j=1}^{p} \{u_{kp+i,lp+j} \, e_{kp+i,lp+j}\}$ for a fixed control function $u(.)$ bounded by B. Consider the Riccati type equation,

$$\frac{dS}{dt} = -A(t)'S - SA(t) + C'r^{-1}C - SQS, \tag{93}$$

$$S(0) = S_0, \tag{94}$$

where Q and r are given $Np \times Np$ and $p \times p$ symmetric positive definite matrices.

We are led to consider also the associated other Riccati type equation:

$$\begin{aligned} \frac{dP}{dt} &= PA(t)' + A(t)P - PC'r^{-1}CP + Q, \\ P(0) &= P_0. \end{aligned} \tag{95}$$

Of course we look for symmetric solutions of both equations.

The following standard facts will be useful in this section.

Facts:

1. A, B are $n \times n$ matrices. $|trace(AB)| \leq \sqrt{trace(A'A)}\sqrt{trace(B'B)}$,
2. S is $n \times n$, symmetric positive semidefinite. $trace(S^2) \leq (trace(S))^2$,
3. S is as in 2, then $trace(S^2) \geq \frac{1}{n}(trace(S))^2$,
4. S is as in 2, then, $trace(SQS) \geq \frac{q}{n}(trace(S))^2$, with $q = \min_{\|x\|=1} x'Qx$,
5. (consequence of 2, 3), S is as in 2, $|.|$ is any norm on $n \times n$ matrices, then

$$l|S| \leq trace(S) \leq m|S|, \tag{96}$$

for $l, m > 0$. In particular, this is true for the Frobenius norm $\|.\|$ associated to the Euclidean norm $\|.\|$ on R^n. (For A, $n \times n$, $\|A\| = (\rho(A'A))^{\frac{1}{2}}$, ρ the spectral radius).

Here, we set $m = Np$, and we denote by S_m the set of symmetric $m \times m$ matrices, and by $S_m(+)$ the set of positive definite matrices. For $S \in S_m$, we set $|S| = \sqrt{trace(S^2)}$.

Lemma 2.12. *For any $\lambda \in R^*$, any solution $S : [0, T[\to S_m$ of (93) (possibly $T = +\infty$), we have, for all $t \in [0, T[$:*

$$S(t) = e^{-\lambda t}\Phi_u(t,0)S_0\Phi_u'(t,0) + \int_0^t e^{-\lambda(t-s)}\Phi_u(t,s)C'r^{-1}C\Phi_u'(t,s)ds$$
$$+\lambda\int_0^t e^{-\lambda(t-s)}\Phi_u(t,s)\left(S(s) - \frac{S(s)QS(s)}{\lambda}\right)\Phi_u'(t,s)ds, \qquad (97)$$

where $\Phi_u(t,s)$ is the solution of $\frac{d\Phi_u(t,s)}{dt} = -A(t)'\Phi_u(t,s)$, $\Phi_u(t,t) = Id_m$.

Proof. Let $\hat{S} : [0, T[\to S_m$, $\hat{S}(t) = e^{\lambda t}S(t)$. Then $\hat{S}(t)$ satisfies the equation

$$\frac{d\hat{S}}{dt}(t) = -A(t)'\hat{S}(t) - \hat{S}(t)A(t) + e^{\lambda t}C'r^{-1}C + \lambda\hat{S}(t) - e^{-\lambda t}\hat{S}(t)Q\hat{S}(t),$$

for $t \in [0, T[$. Applying the variation of constants formula to the equation

$$\frac{d\Lambda}{dt}(t) = -A(t)'\Lambda(t) - \Lambda(t)A(t) + F(t),$$

we get, going back to S, the relation (97). ■

We will need also the following auxiliary lemma:

Lemma 2.13. *Let a, b, and c be three positive constants. Let $x : [0, T[\to R_+$ (possibly $T = +\infty$) be an absolutely continuous function satisfying for almost all t, $0 < t < T$, the inequality*

$$\dot{x} \leq -ax^2 + 2bx + c.$$

Denote by $\xi, -\eta$, the roots of $-aX^2 + 2bX + c = 0$, $\xi, \eta > 0$. Then $x(t) \leq \max(x(0), \xi)$ for all $t \in [0, T[$. If $x(0) > \xi$, then, for all $t > 0$ in $[0, T[$:

$$x(t) \leq \xi + \frac{\xi + \eta}{e^{a(\xi+\eta)t} - 1}.$$

Proof. If $x(t) \leq \xi$ for all $t \in [0, T[$, we are done. Assume now that $E = \{t \in [0, T[\mid x(t) > \xi\} \neq \emptyset$. Take any connected component $]\alpha, \beta[$ of E such that $\alpha > 0$. Either $\beta < T$, hence $x(\alpha) = x(\beta) = \xi$, or $\beta = T$, and $x(\alpha) = \xi$, $x(\beta) \geq \xi$. Hence, in all cases, $x(\alpha) = \xi$, $x(\beta) \geq \xi$. The set F of all $\tau \in]\alpha, \beta[$ such that $\dot{x}(\tau) \geq 0$ has positive measure: otherwise, $x(\beta) < x(\alpha)$. But for

$\tau \in F, \dot{x}(\tau) \leq a(\xi - x(\tau))(x(\tau) + \eta) < 0$, a contradiction. Hence, either $E = \emptyset$ or $E = [0, t_1[, t_1 > 0$. In the first case, $x(t) \leq \xi$ for all $t \in [0, T[$. In the second case, $x(t) \leq \xi$ for $t \geq t_1$. On $[0, t_1[, \dot{x}(\tau) \leq a(\xi - x(\tau))(x(\tau) + \eta) < 0$ a.e., and so, $x(\tau) \leq x(0)$ for $\tau \in [0, t_1[$.

In the second case, on $[0, t_1[$, $\frac{-\dot{x}(\tau)}{a(x(\tau)-\xi)(x(\tau)+\eta)} \geq 1$ a.e., hence,

$$\log \frac{x(\tau) + \eta}{x(\tau) - \xi} \geq a(\xi + \eta)\tau + \log \frac{x(0) + \eta}{x(0) - \xi},$$

for $\tau \in [0, t_1[$.

Therefore, $\frac{x(\tau)+\eta}{x(\tau)-\xi} \geq e^{a(\xi+\eta)\tau}$. This gives $x(\tau) \leq \xi + \frac{\eta+\xi}{e^{a(\xi+\eta)\tau}-1}$ for $0 \leq \tau \leq t_1$. However, because $x(\tau) \leq \xi$ for $\tau \geq t_1$, $x(\tau) \leq \xi + \frac{\eta+\xi}{e^{a(\xi+\eta)\tau}-1}$ for $\tau \in]0, T[$. ■

Lemma 2.14. *If $S : [0, T[\rightarrow S_m(+)$ is a solution of Equation (93), then, for almost all $t \in [0, T[$*

$$\frac{d(trace(S(t))}{dt} \leq -a\ trace(S(t))^2 + 2b\ trace(S(t)) + c,$$

where $a = \frac{1}{m}\lambda_{\min}$, $\lambda_{\min}$ is the smallest eigenvalue of Q, $b = \sup_t \sqrt{trace(A(t)'A(t)}$, $c = trace(C'r^{-1}C)$.

Proof. This follows from (93) and facts 1 to 5 above. ■

Lemma 2.15. *Let $S : [0, e(S)[\rightarrow S_m$ be a maximal semisolution of (93). If $S(0) = S_0$ is positive definite, then $e(S) = +\infty$ and $S(t)$ is positive definite for all $t \geq 0$.*

Proof. Assume that S is not always positive definite. Let $\theta = \inf\{t \mid S(t) \notin S_m(+)\}$. Then $S(t) \in S_m(+)$ for all $t \in [0, \theta[$. Using Lemmas 2.13 and 2.14, we see that $trace(S(t)) \leq \max(trace(S_0), \xi)$. Then, $|S(t)| = \sqrt{trace(S(t)^2)} \leq trace(S(t)) \leq \max(trace(S_0), \xi)$. Choose $\lambda > |Q| \max(trace(S_0), \xi)$. Apply Lemma 2.12 with $t = \theta$. $S(\theta) = (I) + (II) + (III)$,

$$(I) = e^{-\lambda\theta}\Phi_u(\theta, 0)S_0\Phi_u'(\theta, 0),$$

$$(II) = \int_0^\theta e^{-\lambda(\theta-s)}\Phi_u(\theta, s)C'r^{-1}C\Phi_u'(\theta, s)\,ds,$$

$$(III) = \lambda \int_0^\theta e^{-\lambda(\theta-s)}\Phi_u(\theta, s)\left(S(s) - \frac{S(s)QS(s)}{\lambda}\right)\Phi_u'(\theta, s)\,ds.$$

(I) is obviously positive definite. (II) is positive semidefinite at least. For (III):

$$\left(S(s) - \frac{S(s)QS(s)}{\lambda}\right) = \sqrt{S(s)}\left(Id_m - \frac{\sqrt{S(s)}Q\sqrt{S(s)}}{\lambda}\right)\sqrt{S(s)},$$

and $|\sqrt{S(s)}Q\sqrt{S(s)}| \leq |S(s)||Q| \leq |Q| \max(trace(S_0), \xi)$. Hence, $(S(s) - \frac{S(s)QS(s)}{\lambda})$ is positive definite. So is (III). Therefore, $S(\theta)$ is positive definite, a contradiction.

Now, by Lemma 2.14 and Lemma 2.13, $trace(S(t) \leq \max(trace(S_0), \xi)$ for all $t \in [0, e(S)[$. This implies that $|S(t)| \leq trace(S(t) \leq \max(trace(S_0), \xi)$. Hence, $S(t)$ remains bounded as $t \to e(S)$, and $e(S) = +\infty$. ■

Let $S : [0, +\infty[\to S_m$ be a solution of (93), such that $S(0) = S_0 \in S_m(+)$. Then, as we know from Lemma 2.15 , $S(t) \in S_m(+)$ for all $t \in R_+$, and by Lemmas 2.14 and 2.13, $trace(S(t)) \leq \xi + \frac{\xi+\eta}{e^{a(\xi+\eta)t}-1}$. Because $S(t) \leq trace(S(t))\, Id_m$, we get:

Lemma 2.16. *Assume that $S_0 \in S_m(+)$. For all $T_2 > 0$, for all $t \geq T_2$, $S(t) \leq A(T_2)\, Id_m$, with $A(T_2) = (\xi + \frac{\xi+\eta}{e^{a(\xi+\eta)T_2}-1})$.*

Lemma 2.17. *For each time $T_3 > 0$, there exists a constant $\gamma_3 > 0$ such that, if $S : [0, +\infty[\to S_m$ is a solution of (93) such that $S(0) = S_0 \in S_m(+)$, then:*

$$S(t) \geq \gamma_3\, Id_m, \ for\ t \geq T_3.$$

Proof. Chose $\lambda > A\,|Q|$. Apply Lemma 2.12 but starting from $T > 0$. Then for $t \geq T$, $S(t) = (I) + (II) + (III)$,

$$(I) = e^{-\lambda(t-T)}\Phi_u(t, T)S(T)\Phi'_u(t, T),$$

$$(II) = \int_T^t e^{-\lambda(t-s)}\Phi_u(t, s)C'r^{-1}C\Phi'_u(t, s)\, ds,$$

$$(III) = \lambda \int_T^t e^{-\lambda(t-s)}\Phi_u(t, s)\left(S(s) - \frac{S(s)QS(s)}{\lambda}\right)\Phi'_u(t, s)\, ds.$$

(I) is obviously positive definite. If $T \geq T_2$, using a reasoning similar to the one in the proof of Lemma 2.15 and the bound in Lemma 2.16, we see that for all $\sigma \in [T, t]$, $(S(s) - \frac{S(s)QS(s)}{\lambda})$ is positive definite. Hence, (III) is positive definite.

From now on, we take $T = T_2$, the arbitrary constant of Lemma 2.16. For $t \geq T_2 + \hat{T}_0$, $\hat{T}_0$ arbitrary, we have

$$(II) \geq \int_{t-\hat{T}_0}^{t} e^{-\lambda(t-s)}\Phi_u(t,s)C'r^{-1}C\Phi_u'(t,s)\,ds,$$

$$(II) \geq \int_{t-\hat{T}_0}^{t} e^{-\lambda(t-s)}\Phi_{\bar{u}}(\hat{T}_0, s-t+\hat{T}_0)C'r^{-1}C\Phi_{\bar{u}}'(\hat{T}_0, s-t+\hat{T}_0)\,ds,$$

where $\bar{u}$ is the control $\bar{u}(t') = u(t'+t-\hat{T}_0)$. Then

$$(II) \geq \int_{0}^{\hat{T}_0} e^{-\lambda(\hat{T}_0-s)}\Phi_{\bar{u}}(\hat{T}_0, s)C'r^{-1}C\Phi_{\bar{u}}'(\hat{T}_0, s)\,ds,$$

By Lemma 2.11, for all $\hat{T}_0 > 0$, there exists a constant $\gamma_0 > 0$ such that

$$\int_{0}^{\hat{T}_0} \Phi_{\bar{u}}(\hat{T}_0, s)C'r^{-1}C\Phi_{\bar{u}}'(\hat{T}_0, s)\,ds \geq \gamma_0\, Id_m.$$

Then,

$$(II) \geq e^{-\lambda\hat{T}_0}\int_{0}^{\hat{T}_0} \Phi_{\bar{u}}(\hat{T}_0, s)C'r^{-1}C\Phi_{\bar{u}}'(\hat{T}_0, s)\,ds,$$

$$(II) \geq e^{-\lambda\hat{T}_0}\gamma_0\, Id_m.$$

Now, $S(t) \geq (II) \geq e^{-\lambda\hat{T}_0}\gamma_0\, Id_m$, if $t \geq T_2 + \hat{T}_0$. Hence, we get the lemma taking $T_3 = T_2 + \hat{T}_0$, $\gamma_3 = e^{-\lambda\hat{T}_0}\gamma_0$. ■

The following theorem summarizes all the results in the previous lemmas.

Theorem 2.18. *The solution $S(t)$ of the Riccati equation (93) is well defined and positive definite for all $t \geq 0$. Moreover, $S(t) = P(t)^{-1}$, where $P(t)$ is the solution of (95) for $P_0 = (S_0)^{-1}$, and for all $T > 0$, there are constants $0 < \gamma < \delta$, depending on T, B, Q, r only (not on S_0!) such that, for $t \geq T$:*

$$\gamma Id_{Np} \leq S(t) \leq \delta Id_{Np}.$$

One has to be very careful: all original versions of statements and proofs of this theorem are wrong. In particular, the following classical inequality is false:

$$\left(\frac{\alpha_2}{1+\alpha_2\beta_1}\right)Id_{Np} \leq P(t) \leq Id_{Np}\left(\frac{1}{\alpha_1}+\beta_2\right),$$

where α_1, β_1 are the bounds on the Gramm observability matrix (α, β in Lemma 2.11), and α_2, β_2 are the bounds on the Gramm controllability matrix.

For t small, $t \leq T$, we will need a more straightforward estimation:

Lemma 2.19. *There is a function $\varphi(T) = \varphi_1(T) + \|P(0)\|\varphi_2(T)$, depending on $\|P(0)\|$, Q, $s = Sup_{0\leq t\leq T}\|A(t)\|$ only, such that $\|P(t)\| \leq \varphi(T)$, $0 \leq t \leq T$.*

Proof. For $0 \leq t \leq T$, Equation (95) gives, after elementary manipulations,

$$\|P(t)\| \leq \|P(0)\| + \int_0^t (2s\|P(\tau)\| + \|Q\|)\, d\tau,$$

which, using Gronwall's inequality, gives

$$\|P(t)\| \leq (\|P(0)\| + \|Q\|T)e^{2sT}.$$ ■

2.4.3. Proof of Theorems 2.6 and 2.8, and Corollaries 2.7 and 2.9

Recall that $\Delta = BlockDiag(Id_p, \frac{1}{\theta}Id_p, \ldots, (\frac{1}{\theta})^{N-1}Id_p)$. We set

$$\tilde{x} = \Delta x, \tilde{z} = \Delta z, \varepsilon = z - x, \tilde{\varepsilon} = \Delta\varepsilon, \tilde{S} = \theta\Delta^{-1}S\Delta^{-1},$$
$$\tilde{P} = \tilde{S}^{-1} = \tfrac{1}{\theta}\Delta P\Delta, \tilde{b}(z) = \Delta b(\Delta^{-1}z), \tilde{b}^*(z) = \Delta b^*(\Delta^{-1}z)\Delta^{-1}. \tag{98}$$

Exercise 2.5. Show that:

1. $\tilde{b}(z)$ is Lipschitz, with the same Lipschitz constant as the one of b,
2. $\tilde{b}^*$ is bounded, with the same bound as the one of b^* ($\theta \geq 1$).

Trivial computations give the following:

$$\frac{d}{dt}(\tilde{\varepsilon}) = \theta\left[(A_{Np} - \tilde{P}C'r^{-1}C)\tilde{\varepsilon} + \frac{1}{\theta}(\tilde{b}(\tilde{z}) - \tilde{b}(\tilde{x}))\right],$$
$$\frac{d}{dt}(\tilde{S}) = \theta\left[-\left(A'_{Np} + \frac{1}{\theta}\tilde{b}^{*'}\right)\tilde{S} - \tilde{S}\left(A_{Np} + \frac{1}{\theta}\tilde{b}^*\right) + C'r^{-1}C - \tilde{S}Q\tilde{S}\right],$$
$$\frac{d}{dt}(\tilde{P}) = \theta\left[\tilde{P}\left(A'_{Np} + \frac{1}{\theta}\tilde{b}^{*'}\right) + \left(A_{Np} + \frac{1}{\theta}\tilde{b}^*\right)\tilde{P} - \tilde{P}C'r^{-1}C\tilde{P} + Q\right]. \tag{99}$$

Therefore, we set

$$\bar{\varepsilon}\left(\frac{t}{\theta}\right) = \tilde{\varepsilon}(t), \bar{S}\left(\frac{t}{\theta}\right) = \tilde{S}(t), \bar{P}\left(\frac{t}{\theta}\right) = \tilde{P}(t),$$

and we get

$$\frac{d}{dt}(\bar{\varepsilon}) = (A_{Np} - \bar{P}C'r^{-1}C)\bar{\varepsilon} + \frac{1}{\theta}(\tilde{b}(\bar{z}) - \tilde{b}(\bar{x}))],$$
$$\frac{d}{dt}(\bar{S}) = -\left(A'_{Np} + \frac{1}{\theta}\tilde{b}^{*'}(\bar{z})\right)\bar{S} - \bar{S}\left(A_{Np} + \frac{1}{\theta}\tilde{b}^*(\bar{z})\right) + C'r^{-1}C - \bar{S}Q\bar{S},$$
$$\frac{d}{dt}(\bar{P}) = \bar{P}\left(A'_{Np} + \frac{1}{\theta}\tilde{b}^{*'}(\bar{z})\right) + \left(A_{Np} + \frac{1}{\theta}\tilde{b}^*(\bar{z})\right)\bar{P} - \bar{P}C'r^{-1}C\bar{P} + Q. \tag{100}$$

We fix $T > 0$, and we consider successively the two situations $t \leq T$ and $t \geq T$.

Proof of Theorems 2.6, 2.8. The following is a direct consequence of Lemma 2.11 and Theorem 2.18: Because the control u is bounded by B and $\theta > 1$, for $t \geq T$, we can find constants, $0 < \alpha(T) < \beta(T)$, independent of $\bar{S}_0$, such that

$$\begin{aligned} &\alpha(T)Id \leq \bar{S}(t) \leq \beta(T)Id, \\ &\frac{1}{\beta(T)}Id \leq \bar{P}(T) \leq \frac{1}{\alpha(T)}. \end{aligned} \tag{101}$$

It will be important that $\alpha(T)$, $\beta(T)$ do not depend on θ.

Also, by Theorem 2.18, if we start with an initial condition $S(0)$ (resp. $P(0)$) in the set of positive definite symmetric matrices, then, for all $t \geq 0$, $\bar{S}(t)$ (resp. $\bar{P}(t)$) is also positive definite.

Starting from (100), the following computations are straightforward:

$$\begin{aligned} \frac{d}{dt}(\bar{\varepsilon}(t)'\bar{S}(t)\bar{\varepsilon}(t)) &= 2\bar{\varepsilon}(t)'\bar{S}(t)\frac{d}{dt}(\bar{\varepsilon}(t)) + \bar{\varepsilon}(t)'\frac{d}{dt}(\bar{S}(t))\bar{\varepsilon}(t) \\ &= 2\bar{\varepsilon}(t)'\bar{S}(t)\left(A_{Np}\bar{\varepsilon} + \frac{1}{\theta}(\tilde{b}(\bar{z}) - \tilde{b}(\bar{x}))\right) - 2(C\bar{\varepsilon})'r^{-1}C\bar{\varepsilon} \\ &\quad - 2\bar{\varepsilon}(t)'\bar{S}(t)A_{Np}\bar{\varepsilon}(t) - 2\bar{\varepsilon}(t)'\bar{S}(t)\frac{1}{\theta}\tilde{b}^*(\bar{z})\bar{\varepsilon} \\ &\quad + (C\bar{\varepsilon})'r^{-1}C\bar{\varepsilon} - \bar{\varepsilon}(t)'\bar{S}(t)Q\bar{S}(t)\bar{\varepsilon}(t). \end{aligned}$$

If we denote by $Q_m > 0$, the smallest eigenvalue of Q, we have, for $T \leq t$:

$$-\bar{S}(t)Q\bar{S}(t) \leq -Q_m\alpha(T)\bar{S}(t),$$

which implies

$$\begin{aligned} \frac{d}{dt}(\bar{\varepsilon}(t)'\bar{S}(t)\bar{\varepsilon}(t)) &\leq -Q_m\alpha(T)\bar{\varepsilon}'\bar{S}(t)\bar{\varepsilon} + 2\bar{\varepsilon}'\bar{S}(t)\left(\frac{1}{\theta}(\tilde{b}(\bar{z}) - \tilde{b}(\bar{x}) - \tilde{b}^*(\bar{z})\bar{\varepsilon})\right) \\ &\leq -Q_m\alpha(T)\bar{\varepsilon}'\bar{S}(t)\bar{\varepsilon} + \|\bar{S}(t)\|\frac{L}{\theta}\|\bar{\varepsilon}\|^2, \end{aligned}$$

because of Exercise 2.5. Hence,

$$\begin{aligned} \frac{d}{dt}(\bar{\varepsilon}(t)'\bar{S}(t)\bar{\varepsilon}(t)) &\leq -\bar{\varepsilon}'\bar{S}(t)\bar{\varepsilon}\left(Q_m\alpha(T) - \|\bar{S}(t)\|\frac{L}{\theta\alpha(T)}\right) \\ &\leq -\bar{\varepsilon}'\bar{S}(t)\bar{\varepsilon}\left(Q_m\alpha(T) - \frac{\beta(T)}{\alpha(T)}\frac{L}{\theta}\right). \end{aligned}$$

This implies that, for all t, $T \leq t$,

$$\bar{\varepsilon}(t)'\bar{S}(t)\bar{\varepsilon}(t) \leq \bar{\varepsilon}(T)'\bar{S}(T)\bar{\varepsilon}(T)e^{-(Q_m\alpha(T)-\frac{\beta(T)}{\alpha(T)}\frac{L}{\theta})(t-T)}.$$

Therefore, $\bar{\varepsilon}(t)'\bar{S}(t)\bar{\varepsilon}(t) \leq \beta(T)\|\bar{\varepsilon}(T)\|^2 e^{-(Q_m\alpha(T)-\frac{\beta(T)}{\alpha(T)}\frac{L}{\theta})(t-T)}$, and finally,

$$t \geq T: \qquad (102)$$
$$\|\bar{\varepsilon}(t)\|^2 \leq \frac{\beta(T)}{\alpha(T)}e^{-(Q_m\alpha(T)-\frac{\beta(T)}{\alpha(T)}\frac{L}{\theta})(t-T)}\|\bar{\varepsilon}(T)\|^2.$$

It follows that, for $\tau \geq \frac{T}{\theta}$,

$$\|\tilde{\varepsilon}(\tau)\|^2 \leq \frac{\beta(T)}{\alpha(T)}e^{-(Q_m\alpha(T)\theta-\frac{\beta(T)}{\alpha(T)}L)(\tau-\frac{T}{\theta})}\left\|\tilde{\varepsilon}\left(\frac{T}{\theta}\right)\right\|^2,$$

and

$$\|\varepsilon(\tau)\|^2 \leq \theta^{2(N-1)}\frac{\beta(T)}{\alpha(T)}e^{-(Q_m\alpha(T)\theta-\frac{\beta(T)}{\alpha(T)}L)(\tau-\frac{T}{\theta})}\left\|\varepsilon\left(\frac{T}{\theta}\right)\right\|^2,$$

which proves Theorems 2.6 and 2.8. ■

Proof of Corollaries 2.7 and 2.9. Starting from (102), and using the result of Lemma 2.19, it is only a matter of trivial computations to prove the corollaries. Let us give the details: We consider the case $t \leq T$, and we assume that $S(0) = S_0$ lies in the compact set: $cId \leq S_0 \leq dId$. As a consequence, $P(0) \leq \frac{1}{c}Id$. By Equation (100), we have, for $t \leq T$: $\frac{d}{dt}(\bar{\varepsilon}) = (A_{Np} - \bar{P}C'r^{-1}C)\bar{\varepsilon} + \frac{1}{\theta}(\tilde{b}(\bar{z}) - \tilde{b}(\bar{x}))$, hence,

$$\|\bar{\varepsilon}(t)\|^2 \leq \|\bar{\varepsilon}(0)\|^2 + \int_0^t \|\bar{\varepsilon}(\tau)\|^2\left(2\|A_{Np}\| + 2\|C\|^2\|r^{-1}\|\ \|\bar{P}\| + \frac{\nu}{\theta}\right)d\tau,$$

and by Lemma 2.19, we know that $\|\bar{P}(t)\| \leq \varphi_1(T) + \|\bar{P}_0\|\varphi_2(T)$. Then, because $\bar{P}_0 = \frac{1}{\theta}\Delta P_0\Delta$, $\theta > 1$, $\|\bar{P}(t)\| \leq \varphi_1(T) + \|P_0\|\varphi_2(T) \leq \varphi_1(T) + \frac{1}{c}\varphi_2(T) = \varphi(T)$.

$$\|\bar{\varepsilon}(t)\|^2 \leq \|\bar{\varepsilon}(0)\|^2 + (2s + 2\|C\|^2\|r^{-1}\|\ \varphi(T) + \nu)\int_0^t \|\bar{\varepsilon}(\tau)\|^2 d\tau.$$

Gronwall's inequality implies that

$$\|\bar{\varepsilon}(t)\|^2 \leq \Psi(T)\|\bar{\varepsilon}(0)\|^2,$$

with $\Psi(T) = e^{(2s+2\|C\|^2\|r^{-1}\|\ \varphi(T)+\nu)T}$. In particular, $\|\bar{\varepsilon}(T)\|^2 \leq \Psi(T)\|\bar{\varepsilon}(0)\|^2$.

Plugging this in (102), we get

$$\|\bar{\varepsilon}(t)\|^2 \leq \frac{\beta(T)}{\alpha(T)} e^{-(Q_m\alpha(T)-\frac{\beta(T)}{\alpha(T)}\frac{L}{\theta})(t-T)} \Psi(T)\|\bar{\varepsilon}(0)\|^2, \text{ for } t \geq T. \quad (103)$$

Going back to $t \leq T$, we have

$$\|\bar{\varepsilon}(t)\|^2 \leq \Psi(T)\|\bar{\varepsilon}(0)\|^2 \leq \Psi(T)\frac{\beta(T)}{\alpha(T)}\|\bar{\varepsilon}(0)\|^2$$
$$\leq \frac{\beta(T)}{\alpha(T)} e^{(Q_m\alpha(T)-\frac{\beta(T)}{\alpha(T)}\frac{L}{\theta})(T-t)} \Psi(T)\|\bar{\varepsilon}(0)\|^2,$$

as soon as $(Q_m\alpha(T) - \frac{\beta(T)}{\alpha(T)}\frac{L}{\theta}) > 0$, i.e., $\theta > \frac{\beta(T)L}{Q_m\alpha(T)^2} = \theta_0(T)$. Hence, in all cases (either $t \leq T$ or $T \leq t$), if $\theta \geq \theta_0(T)$, we have

$$\|\bar{\varepsilon}(t)\|^2 \leq \frac{\beta(T)}{\alpha(T)} e^{-(Q_m\alpha(T)-\frac{\beta(T)}{\alpha(T)}\frac{L}{\theta})(t-T)} \Psi(T)\|\bar{\varepsilon}(0)\|^2.$$

This can be rewritten as

$$\|\bar{\varepsilon}(t)\|^2 \leq \frac{\beta(T)}{\alpha(T)} e^{(Q_m\alpha(T)-\frac{\beta(T)}{\alpha(T)}\frac{L}{\theta})T} e^{-(Q_m\alpha(T)-\frac{\beta(T)}{\alpha(T)}\frac{L}{\theta})t} \Psi(T)\|\bar{\varepsilon}(0)\|^2,$$

and

$$\|\bar{\varepsilon}(t)\|^2 \leq H(T)e^{-(Q_m\alpha(T)-\frac{\beta(T)}{\alpha(T)}\frac{L}{\theta})t}\|\bar{\varepsilon}(0)\|^2,$$

with $H(T) = \frac{\beta(T)}{\alpha(T)}\Psi(T)e^{Q_m\alpha(T)T}$. Therefore, setting $t = \theta\tau$:

$$\|\tilde{\varepsilon}(\tau)\|^2 \leq H(T)e^{-(Q_m\alpha(T)\theta-\frac{\beta(T)}{\alpha(T)}L)\tau}\|\tilde{\varepsilon}(0)\|^2, \tau \geq 0.$$

Because $\varepsilon = \Delta^{-1}\tilde{\varepsilon}$, and $\theta > 1$, $\|\varepsilon(\tau)\|^2 \leq \|(\Delta^{-1})\|^2\|\tilde{\varepsilon}(\tau)\|^2 \leq \theta^{2(N-1)}\|\tilde{\varepsilon}(\tau)\|^2$, we get, for all $t \geq 0$:

$$\|\varepsilon(t)\|^2 \leq \theta^{2(N-1)}H(T)e^{-(Q_m\alpha(T)\theta-\frac{\beta(T)}{\alpha(T)}L)t}\|\varepsilon(0)\|^2.$$

Corollaries 2.7 and 2.9 are proven. ■

Remark 2.1. Note that the function $H(T)$ depends on c, $S(0) \geq cId$. So that, contrarily to the case of the high-gain Luenberger observer, the function $k(.)$ with polynomial growth in the definitions (1.2, 1.4) of exponential observers actually depends on the compact $\bar{K}$ in the definitions.

2.4.4. The Continuous-Discrete Version of the High-Gain Extended Kalman Filter

This is a more realistic version of the previous high gain observer: Observations are sampled. As the continuous high-gain extended Kalman filter, it applies to systems that are in the normal form (90), in restriction to compact sets. In particular, it applies to all systems that have a phase-variable representation for sufficiently smooth controls, and to control affine systems that have a uniform canonical flag, for general L^∞ controls.

For the statement of our result, let us make exactly the same assumptions as in Section 2.4.1. For simplicity in exposition, let us consider the **single output** case only.

Let us choose a time step δt that is small enough. The equations of the continuous-discrete version $\Sigma_{Oc.d.}$ of our extended Kalman filter are, for $t \in [(k-1)\delta t, k\delta t[$:

(Prediction step)

$$\begin{aligned} &(i)\ \frac{dz}{dt} = Az + b(z,u), \\ &(ii)\ \frac{dS}{dt} = -(A + b^*(z,u))'S - S(A + b^*(z,u)) - SQ_\theta S, \end{aligned} \tag{104}$$

and at time $k\delta t$:

(Innovation step)

$$\begin{aligned} &(i)\ z_k(+) = z_k(-) - S_k(+)^{-1}C'r^{-1}\delta t(Cz_k(-) - y_k), \\ &(ii)\ S_k(+) = S_k(-) + C'r^{-1}C\delta t. \end{aligned} \tag{105}$$

The assumptions being the same as for Theorem 2.6 and Corollary 2.7, we have

Theorem 2.20. *For all $T > 0$, there are two positive constants, θ_0 and μ, such that, for all δt small enough, $\theta > \theta_0$ and $\theta\ \delta t < \mu$, one has, for all $t > T$:*

$$\|z(t) - x(t)\| \le \bar{k}\theta^{n-1}e^{-(\lambda\theta-\omega)(t-\frac{T}{\theta})}\left\|z\left(\frac{T}{\theta}\right) - x\left(\frac{T}{\theta}\right)\right\|,$$

for some positive constants $\bar{k}$, λ, ω.

This is the continuous-discrete analog of Formula (89). Hence, it is possible to state the continuous-discrete analogs of the other corollaries in the previous section. We leave this to the reader.

Lemma 2.21. *If Q is symmetric positive definite, and if λ is small, then*

$$(Q + \lambda QC'CQ)^{-1} = Q^{-1} - C'(\lambda^{-1} + CQC')^{-1}C.$$

Exercise 2.6. Prove Lemma 2.21.

Proof of Theorem 2.20. We only sketch the proof. All details are similar to those of the proof of Theorem 2.8 and the notations are the same. For details of these computations in the standard linear case, see [29].

For $t \in [(k-1)\delta t, k\delta t[$,

$$(i)\ \frac{d}{dt}(\tilde{\varepsilon}'\tilde{S}\tilde{\varepsilon}) \leq \theta\left[-\tilde{\varepsilon}'\tilde{S}Q\tilde{S}\tilde{\varepsilon} + 2\tilde{\varepsilon}'\tilde{S}\left(\frac{1}{\theta}(\tilde{b}(\tilde{z}) - \tilde{b}(\tilde{x}) - \tilde{b}^*(\tilde{z})\tilde{\varepsilon})\right)\right],$$
$$(ii)\ \frac{d}{dt}\tilde{S} = \theta\left(-\tilde{S}Q\tilde{S} - \left(A' + \frac{1}{\theta}\tilde{b}^{*\prime}\right)\tilde{S} - \tilde{S}\left(A + \frac{1}{\theta}\tilde{b}^*\right)\right), \tag{106}$$

at $k\delta t$, we use Lemma 2.21 to get

$$(i)\ (\tilde{\varepsilon}'\tilde{S}\tilde{\varepsilon})_k(+) = (\tilde{\varepsilon}'\tilde{S}\tilde{\varepsilon})_k(-) - \tilde{\varepsilon}_k(-)'C'\left(C\tilde{S}_k(-)^{-1}C' + \frac{r}{\theta\delta t}\right)^{-1} C\tilde{\varepsilon}_k(-),$$
$$(ii)\ \tilde{S}_k(+) = \tilde{S}_k(-) + C'r^{-1}C\theta\delta t, \tag{107}$$

provided that $\theta\delta t$ is small, $\theta\delta t < \mu$.

By (107), (i), $(\tilde{\varepsilon}'\tilde{S}\tilde{\varepsilon})_k(+) \leq (\tilde{\varepsilon}'\tilde{S}\tilde{\varepsilon})_k(-)$. We set

$$B(\tilde{S}) = \|\tilde{S}\|\ \|\tilde{S}^{-1}\|,\ B_k = \sup_{t\in[(k-1)\delta t, k\delta t[} B(\tilde{S}(t)),$$

$$\lambda_k = \sup_{t\in[(k-1)\delta t, k\delta t[} \frac{\lambda_{\min}(\tilde{S}Q\tilde{S})}{\lambda_{\max}(\tilde{S})}(t),$$

where $\lambda_{\min}$ and $\lambda_{\max}$ denote the smallest and largest eigenvalues.

Simple computations give

$$k \geq 1,$$
$$(\tilde{\varepsilon}'\tilde{S}\tilde{\varepsilon})_k(+) \leq e^{-(\lambda_k\theta - \omega B_k)\delta t}(\tilde{\varepsilon}'\tilde{S}\tilde{\varepsilon})_{k-1}(+). \tag{108}$$

Therefore, to finish, we have to bound $\tilde{S}_k$ from above and from below. Setting again $\bar{S}(t) = \tilde{S}(\frac{t}{\theta})$, we get, for $t \in [\theta(k-1)\delta t, \theta k\delta t[$:

$$\frac{d}{dt}\bar{S} = -\bar{S}Q\bar{S} - \left(A' + \frac{1}{\theta}\tilde{b}^{*\prime}\right)\bar{S} - \bar{S}\left(A + \frac{1}{\theta}\tilde{b}^*\right). \tag{109}$$

At $k\delta t$:

$$\bar{S}_k(+) = \bar{S}_k(-) + C'r^{-1}C\theta\delta t. \tag{110}$$

A reasoning similar to that in the previous section shows that, for all $T > 0$, if δt is small enough, there is $\theta_0 > 0$, such that for $\theta_0 \leq \theta \leq \frac{\mu}{\delta t}$ and for $k\ \delta t \geq T$,

$$\alpha Id \leq \bar{S}_k \leq \beta Id,$$

for some constants $0 < \alpha < \beta$. This, with (108), gives the result. ■

3. Appendix

3.1. Continuity of Input-State Mappings

Let (Σ) be a control affine system on a manifold M:

$$\dot{x} = X(x) + \sum_{i=1}^{d_u} Y_i(x) u_i,$$
$$x(0) = x_0.$$

Here, we consider $T > 0$ fixed, and the space of control functions is $L^{\infty}_{([0,T];R^{d_u})}$ with the weak-* topology. The space of state trajectories of the system is the space $C^0_{([0,T];M)}$ of continuous functions: $P : [0, T] \to M$, with the topology of uniform convergence. We want to prove the following result:

Theorem 3.1. *The input-state mapping $P : u(.) \to x(.)$ has open domain and is continuous for these topologies, on any bounded set.*

As usual in this type of reasoning, we can assume that (1) $M = R^n$, and (2) X and Y_i are compactly supported vector fields. Also, to simplify the computations, let us prove the result for a system with a single control only.

To prove the continuity on a bounded set of the mapping P, it is sufficient to prove that it is sequentially continuous for the weak-* topology, because the restriction to any bounded set of this topology is metrizable.

We fix u, and we consider a sequence (u_n), converging *-weakly to u. The corresponding x trajectories are denoted by $x(.)$ and $x_n(.)$ respectively. The following lemma is an obvious consequence of Gronwall's inequality and of the fact that X and Y are Lipschitz:

Lemma 3.2. *There is a $k > 0$ such that, $\forall n, \forall t \in [0, T]$, $\|x_n(t) - x(t)\| \leq k \sup_{\theta\in[0,T]} \| \int_0^{\theta} (u_n(s) - u(s)) Y(x(s))\, ds \|$.*

Proof.

$$\|x_n(t) - x(t)\| \leq \left\| \int_0^t [X(x_n) + Y(x_n)u_n(s) - (X(x) + Y(x)u(s))]\, ds \right\|$$

$$\leq A + B = \left\| \int_0^t [X(x_n) + Y(x_n)u_n(s) - (X(x) + Y(x)u_n(s))]\, ds \right\| + \left\| \int_0^t [Y(x)u_n(s) - Y(x)u(s)]\, ds \right\|.$$

$$B \leq \sup_{\theta\in[0,T]} \| \int_0^\theta (u_n(s) - u(s))Y(x(s))\, ds \|.$$

$$A \leq \int_0^t \|X(x_n) - X(x)\|\, ds + \int_0^t \|Y(x_n) - Y(x)\| |u_n(s)|\, ds.$$

Because u_n converges *-weakly, the sequence $(\|u_n\|_\infty;\, n \in \mathcal{N})$ is bounded. X and Y are Lipschitz, hence,

$A \leq m \int_0^t \|x_n(s) - x(s)\|| \, ds$. Therefore,

$$\|x_n(t) - x(t)\| \leq B + m \int_0^t \|x_n(s) - x(s)\|\, ds.$$

Gronwall's inequality gives the result. ∎

Proof of Theorem 3.1. For all $\theta \in [0, T]$, for all $\delta > 0$, let us consider a subdivision $\{t_j\}$ of $[0, T]$, such that $\theta \in [t_i, t_{i+1}]$ and $t_{j+1} - t_j < \delta$ for all j. We have

$$\left\| \int_0^\theta (u_n(s) - u(s))Y(x(s))\, ds \right\| \leq \left\| \int_0^{t_i} (u_n(s) - u(s))Y(x(s))\, ds \right\| + \left\| \int_{t_i}^\theta (u_n(s) - u(s))Y(x(s))\, ds \right\|.$$

With $\varepsilon > 0$ being given, we take $\delta \leq \frac{\varepsilon}{2\gamma m}$, where $\gamma = \sup_{x\in R^n} \|Y(x)\|$, and $m = \sup_n \|u_n - u\|_\infty$. We get, for all $\theta \in [0, T]$:

$$\left\| \int_0^\theta (u_n(s) - u(s))Y(x(s))\, ds \right\| \leq \left\| \int_0^{t_i} (u_n(s) - u(s))Y(x(s))\, ds \right\| + \frac{\varepsilon}{2}.$$

By the *-weak convergence of (u_n), for $n > N$, we have

$$\left\| \int_0^{t_i} (u_n(s) - u(s))Y(x(s))\, ds \right\| \leq \frac{\varepsilon}{2}.$$

By the lemma, $\forall n > N, \forall t \in [0, T], \|x_n(t) - x(t)\| \leq \varepsilon$. ∎

Part II

Dynamic Output Stabilization and Applications

7
Dynamic Output Stabilization

Using the results of the previous chapters, we can derive a constructive method to solve the following problem. We are given a system Σ:

$$(\Sigma)\ \begin{cases} \dot{x} = f(x, u), \\ y = h(x, u), \end{cases}$$

where $u \in U = I^{d_u}$, and where I is a closed interval of R (possibly unbounded). This system is assumed to have an equilibrium point x_0, which is asymptotically stabilizable by smooth, U-valued state feedback, that is, there is a smooth function $\alpha(x)$, such that x_0 is an asymptotically stable equilibrium of the vector field $\dot{x} = f(x, \alpha(x))$.

The problem is the following: Is it possible to stabilize asymptotically by using not the state information (as does the state feedback $\alpha(x)$), but by using only the output information? As usual in this type of problem, we avoid differentiating the outputs, because, from the physical point of view, we would have to differentiate the noise, or the measurement errors, which is not reasonable.

We will be interested only in the behavior of Σ within the basin of attraction of the equilibrium x_0, hence, we can restrict X to this basin of attraction and assume that $X = R^n$ (see [51]).

The basic idea, coming from the linear theory, is to construct a state observer, and to control the system using the feedback α evaluated on the estimate $\hat{x}$ of the state.

1. We will show that this is possible for the systems of Chapter 3, for which a uniform canonical flag exists, and we can construct an exponential state observer, by the results of the previous chapter.
2. In the general case where we have a phase-variable representation only, i.e., for systems of Chapters 4 and 5, we will show that this is possible by using exponential output observers, but the construction is a bit more sophisticated.

In fact, we will not be able to cover (as in the linear case) the whole original basin of attraction: We will obtain asymptotic stability **within arbitrarily large compact sets** contained in this basin of attraction, only.

At the end of the chapter, we will say a few words about a situation in which the results can be improved from a practical point of view: Our theorems depend very essentially on the high-gain construction. Moreover, the output stabilization is "twice high gain" in a sense that will be clear later on. First, we need high gain for the observer to estimate exponentially. Second, this exponential rate of the observer has to be very large. It is important to understand that in some situations that are very common in practice, only exponential convergence of the estimation of the state is required but the exponential rate can be small.

1. The Case of a Uniform Canonical Flag

We will make the assumptions of Section 2.2 of Chapter 6: $X = R^n$, and our system is globally in the normal form (20). We know already that we can modify the normal form outside any open ball B^0 for the Lipschitz conditions $(A1)$ and $(A2)$ of Chapter 6 to be satisfied globally over X. In the proof of the main theorem (1.1) below, the semitrajectories of Σ under consideration will not leave a compact subset, denoted below by D_{m+1}. This justifies making these assumptions.

We will consider first the **Luenberger type** observer. The same task can be performed by the **Kalman type** observer, but it applies to the control affine case only, and the proof is more complicated, as we shall see.

1.1. Semi-Global Asymptotic Stabilizability

The most convenient notion to be handled in this chapter is not **global asymptotic stabilizability**, but the weaker **semi-global asymptotic stabilizability**, that we define immediately.

Notation. In order to shorten certain statements, let us say that a vector field on X is "**asymptotically stable at $x_0 \in X$ within a compact set $\Gamma \subset X$**" if x_0 is an asymptotically stable equilibrium point and the basin of attraction of x_0 contains Γ.

Definition 1.1. We say that the (unobserved) system Σ on X, $\dot{x} = f(x, u)$ is semiglobally asymptotically stabilizable at (x_0, u_0) if, for each compact $\Gamma \subset X$, there is a smooth feedback $\alpha_\Gamma : X \to Int(U)$, $\alpha_\Gamma(x_0) = u_0$, such that the vector field on X,

$$\dot{x} = f(x, \alpha_\Gamma(x)), \tag{111}$$

is asymptotically stable at x_0 within Γ.

We make the assumption that $X = R^n$, but it may not be clear that the state space X of a semiglobally asymptotically stabilizable Σ should be R^n. However, in fact, this is true: By definition, any compact subset $\Gamma \subset X$ is contained in the basin of attraction of x_0 for a certain smooth vector field. By the results of [51], such a basin of attraction is diffeomorphic to R^n. Then, because we assumed X paracompact, the Brown–Stallings theorem (see [41]) gives the result.

Comment. It does not follow immediately from [51] that the basin of attraction of the origin, for an asymptotically stable vector field, is diffeomorphic to R^n, because it is assumed in the paper that the state space is R^n. However, one can modify slightly the arguments to make them work for a general (paracompact) manifold.

1.2. Stabilization with the Luenberger-Type Observer

We assume (reparametrization) that $x = 0$, $u = 0$ is the equilibrium point under consideration, and the smooth stabilizing feedbacks are $\alpha_\Gamma(x)$, i.e., $\alpha_\Gamma(0) = 0$, and $x = 0$ is an asymptotically stable (within Γ) equilibrium point of the vector field

$$\dot{x} = f(x, \alpha_\Gamma(x)). \tag{112}$$

Now Γ is fixed, together with the corresponding α_Γ. The basin of attraction of zero is B_Γ, $\Gamma \subset B_\Gamma$. We have to study the following system on $X \times X$:

$$\begin{aligned} \dot{x} &= f(x, \tilde{\alpha}(z)), \\ \dot{z} &= f(z, \tilde{\alpha}(z)) - K_\theta(h(z, \tilde{\alpha}(z)) - y), \\ y &= h(x, \tilde{\alpha}(z)), \end{aligned} \tag{113}$$

where K_θ is as in Theorem 2.2 of Chapter 6. The purpose is to show that, if θ is large enough, (0, 0) is an asymptotically stable equilibrium of (113), and the basin of attraction of (0, 0) can be made arbitrarily large in $B_\Gamma \times X$ by increasing θ. Here, $\tilde{\alpha}$ is a certain other smooth feedback, depending on the compact Γ, that we construct now:

Using the inverse Lyapunov theorems (Appendix, Section 3.3), we can find a smooth, proper strict Lyapunov function $V : B_\Gamma \to R_+$, for the vector field

(112), $V(0) = 0$. This means that, along the trajectories of Σ, $\frac{dV(x)}{dt} < 0$ (except for $x = 0$). The function V reaches its maximum m over Γ. Let us consider $D_{m+1} = \{x | V(x) \leq m + 1\}$, and let us replace α_Γ by $\tilde{\alpha}$ such that:

1. $\tilde{\alpha} = \alpha_\Gamma$ on D_{m+1}.
2. $\tilde{\alpha}$ is smooth, compactly supported, with values in $Int(U)$ (we have already assumed above that, by translation in the U space, $0 \in Int(U)$).

Let us also set $B = \sup_{x \in R^n} \|\tilde{\alpha}(x)\|$. We will show the following.

Theorem 1.1. *Given arbitrary compact sets $\Gamma, \Gamma' \subset X$, if θ is large enough, then (113) is asymptotically stable at the origin within $\Gamma \times \Gamma'$.*

Comment. This theorem means that provided that we know a compact set Γ where the system starts, and provided that we modify the stabilizing feedback at infinity, we can just plug the estimate of the state given by our "state observer" into the feedback, and the resulting system is asymptotically stable at the origin. Moreover, the semitrajectories starting from $\Gamma \times \Gamma'$ tend to the origin. That is, **we can asymptotically stabilize at the origin via the observer, using observations only.** This can be done **within arbitrarily large compact sets**.

As we shall see in the proof, it is very important that the observer is exponential, with arbitrary exponential decay: We need the exponential decay for local asymptotic stabilization, but we also need an arbitrarily large exponential in order to estimate the state very quickly.

Proof of Theorem 1.1. The proof is divided into three steps. In this type of considerations, it is always convenient to rewrite the equations of the global dynamics system-observer (113) in terms of the estimate z and the estimation error $\varepsilon = z - x$:

$$\begin{aligned}
\dot{z} &= f(z, \tilde{\alpha}(z)) - K_\theta(h(z, \tilde{\alpha}(z)) - h(z - \varepsilon, \tilde{\alpha}(z))), \\
\dot{\varepsilon} &= f(z, \tilde{\alpha}(z)) - f(z - \varepsilon, \tilde{\alpha}(z)) - K_\theta(h(z, \tilde{\alpha}(z)) - h(z - \varepsilon, \tilde{\alpha}(z))).
\end{aligned} \tag{114}$$

First Step: (Asymptotic Stability of the Origin)

The submanifold $\{\varepsilon = 0\}$ is an invariant manifold for the system (114). Moreover, the dynamics on $\{\varepsilon = 0\}$ is exactly (at least on a neighborhood of zero), the dynamics defined by the vector field (112). Therefore, the dynamics on this invariant manifold is asymptotically stable. On the other hand, let us consider the linearized dynamics at $(z, \varepsilon) = 0$, setting $C = \frac{\partial h}{\partial x}(0, 0)$:

$$\begin{aligned} &(i)\ \dot{z} = \left(\frac{\partial}{\partial z}(f(z, \tilde{\alpha}(z))_{|z=0}\right)z - K_\theta C\varepsilon, \\ &(ii)\ \dot{\varepsilon} = \left(\frac{\partial f}{\partial z}(0, 0)\right)\varepsilon - K_\theta C\varepsilon. \end{aligned} \tag{115}$$

The linear vector field

$$\dot{z} = \left(\frac{\partial}{\partial z}(f(z, \tilde{\alpha}(z))_{|z=0}\right)z$$

has a number n_s (possibly 0) of eigenvalues with strictly negative real part.

Exercise 1.1. Show that the linear vector field on R^n defined by Equation (115, ii) is asymptotically stable provided that θ is large enough.

Hint: Use the same type of time rescaling as in the proof of Theorem 2.2, Chapter 6.

Because of the last exercise, the number of eigenvalues with strictly negative real part of (115) is exactly $n + n_s$. If $n_s = n$, then the asymptotic stability of (114) at the origin follows from the asymptotic stability of the linearized equation (115).

If $n_s < n$, then the remaining $n - n_s$ eigenvalues of the linearized system (115) have zero real part, because (112) is asymptotically stable. In fact, there is a $n - n_s$ dimensional asymptotically stable center manifold $\mathcal{C}$ for the vector field (112) at the origin. By the fact that $\{\varepsilon = 0\}$ is an invariant manifold for (114), it follows that $\mathcal{C} \times \{\varepsilon = 0\}$ is also a (asymptotically stable) center manifold for (114). Hence, by the standard theorems on center manifolds, (114) is asymptotically stable at the origin, provided that θ is large enough.

Second Step: Semitrajectories $(x(t), z(t))$, *starting from* $\Gamma \times \Gamma'$ *are bounded* (Γ' *is an arbitrary compact subset of* R^n)

The dynamics of x is given by:

$$\dot{x} = f(x, \tilde{\alpha}(z)). \tag{116}$$

Over D_{m+1}, $\|\dot{x}\|$ is bounded (D_{m+1} is compact, $f(x, u)$ is smooth, $\tilde{\alpha}(z)$ is bounded because $\tilde{\alpha}$ is compactly supported over R^n). Therefore, all semitrajectories starting from $\Gamma \subset D_m$ need a certain strictly positive time $T_{\min}$ to leave D_{m+1}. Because $\tilde{\alpha}(z)$ is bounded ($\|\tilde{\alpha}(z)\| \leq B$), Theorem 2.2 of Chapter 6 says that the inequality (79) in this theorem is satisfied.

Here is the place where the Lipschitz assumptions (A_1), (A_2) are important. If we have not (A_1), (A_2), but only the normal form (20), then by Exercise 2.1 in Chapter 6, we can modify f outside D_{m+1} for (A_1), (A_2) to hold, and in this case, the inequality (79) holds as long as the semitrajectory $x(t)$ stays in D_{m+1}.

The quantity $\|\hat{x}_0 - x_0\|$ in (79) (equal to $\|z_0 - x_0\|$) is bounded because $(x_0, z_0) \in \Gamma \times \Gamma'$. Therefore, because of the polynomial $k(a)$ against the exponential e^{-at} in this theorem, we can chose θ large enough so that from $T_{\min}$ on, the error $\varepsilon(t)$ becomes smaller than an arbitrary given $\eta > 0$.

Let us consider the compact boundary ∂D_{m+1}. It is a level set of the Lyapunov function V defined above, for the vector field (112): $\dot{x} = f(x, \alpha_\Gamma(x)) = F(x)$. Therefore, on ∂D_{m+1}, $\dot{V}(t) = L_F V(x)$ is smaller than a certain $\mu < 0$. It follows that, if η is small enough, $\dot{V}(t)$ is also negative along the trajectories of $\dot{x} = f(x, \tilde{\alpha}(x + \varepsilon))$, at points of ∂D_{m+1}: $\dot{V}(t) = \frac{\partial V}{\partial x} f(x, \tilde{\alpha}(x + \varepsilon)) = \frac{\partial V}{\partial x} f(x, \alpha_\Gamma(x + \varepsilon))$.

The x-trajectories in (113) cannot leave D_{m+1}, provided that θ is large enough, because ε is small and so $\dot{V}(t)$ remains < 0 on the boundary ∂D_{m+1}.

Comment. At this point, it is important to understand that the stabilization is in a sense "doubly high gain". First, θ has to be large for the observer to perform its task. Second, θ has to be large enough for the trajectories of x stay in D_{m+1} during this time. In many situations we meet in practice (for example in the first application in the next chapter), the trajectories of x are naturally bounded. In that case, the decay of ε has to be exponential, but it is not necessary that this exponential rate is large.

Third Step: Semitrajectories That Stay in $D_{m+1} \times X$ Tend to Zero

This will prove that the basin of attraction of the origin in (113) contains $\Gamma \times \Gamma'$. Take such a semitrajectory $(z(t), \varepsilon(t))$, $t \geq 0$ of (114). This semitrajectory being bounded, its ω-limit set is nonempty. Let (z^*, ε^*) be in this ω-limit set Ω. Then, the whole semitrajectory $(z^*(t), \varepsilon^*(t))$, $t \geq 0$ starting from (z^*, ε^*) is entirely contained in Ω because Ω is invariant. However, because $\varepsilon(t)$ tends to zero when $t \to +\infty$, $\varepsilon^* = 0$. Therefore, $\varepsilon^*(t) = 0$ for all $t \geq 0$. On the other hand, the dynamics on $D_{m+1} \times \{\varepsilon = 0\}$ is just (112), which is asymptotically stable at the origin within D_{m+1}, by assumption (and $D_{m+1} \times \{\varepsilon = 0\}$ is positively invariant by construction). Hence, $(z^*(t), \varepsilon^*(t)) \to 0$. Therefore, the origin is also in the ω-limit set of the semitrajectory $(z(t), \varepsilon(t))$, $t \geq 0$ because the ω-limit set is closed. It follows that the semitrajectory $(z(t), \varepsilon(t))$, $t \geq 0$ enters in a finite time in the basin of attraction of the origin. ■

1.3. Stabilization with the High-Gain EKF

Assume that we are in the control affine case. Then, we could try to use the high-gain extended Kalman filter exactly in the same way (see Theorem 2.6 and Corollary 2.7 in Chapter 6). **There are several additional difficulties.**

Let us consider the systems in normal form (86). Let us set again $\varepsilon = z - x$. The state (z, ε, S) of the couple (system, observer) lives in $R^n \times R^n \times S_n(+)$, where $S_n(+)$ is the set of positive definite symmetric matrices.

$$\begin{aligned} \dot{x} &= Ax + b(x, u), \\ \frac{dz}{dt} &= Az + b(z, u) - S(t)^{-1}C'r^{-1}(Cz - y(t)), \\ \frac{dS}{dt} &= -(A + b^*(z, u))'S - S(A + b^*(z, u)) + C'r^{-1}C - SQ_\theta S, \end{aligned} \tag{117}$$

with

$$u = \tilde{\alpha}(z). \tag{118}$$

As soon as $x(t)$ stays in D_{m+1}, we know that we have the same type of exponential stability as in the previous case (Corollary 2.7 of Chapter 6). Hence, the same argument works, by increasing θ, to show that $x(t)$ stays in D_{m+1} for $t \geq 0$. We also know that $S(t)$ and $S(t)^{-1}$ remain **bounded in the open cone of positive definite matrices** (see (101)).

Hence, all semitrajectories $(z(t), \varepsilon(t), S(t))$, $t \geq 0$ starting from $\Gamma \times \Gamma' \times \Gamma''$ are relatively compact in $R^n \times R^n \times S_n(+)$, where Γ'' is any compact set in $S_n(+)$ (provided that θ is large enough). Moreover, $\varepsilon(t) \to 0$. Hence, a point $(z^*, \varepsilon^*, S^*)$ in the ω-limit set Ω of the given semitrajectory satisfies $\varepsilon^* = 0$. If $\mathcal{T}(t) = (z^*(t), 0, S^*(t))$, $t \geq 0$ is the semitrajectory starting from $(z^*, \varepsilon^*, S^*)$, then again, the equation for $z^*(t)$ is the one in (112), which is asymptotically stable within D_{m+1}, by assumption. Hence, by the invariance and closedness of the ω-limit set, we can take $z^* = 0$. Again, the set $\Xi = \{z = 0, \varepsilon = 0\}$ is invariant under the global dynamics, and the semitrajectory $\check{S}(t)$ of S starting from a point of Ξ is a solution of a time-invariant Riccati equation, relative to an observable time-invariant linear system. It is known from the linear theory that $\check{S}(t)$ tends to the solution S_∞ of the corresponding algebraic Riccati equation. (see [8, 29]). Hence, the point $(0, 0, S_\infty)$ belongs to Ω. It remains to prove the local asymptotic stability of this point.

If we can prove that the linearization at S_∞ of the time-invariant Riccati equation is asymptotically stable, then the result will follow from the same argument as in the proof above: $\{\varepsilon = 0\}$ will be an invariant manifold, containing the asymptotically stable center manifold (if any), and all other eigenvalues of the linearized system will have negative real parts. This exponential stability of a time-invariant Riccati equation also follows from the linear theory. (See for instance [8], Theorem 5.4, p. 73.) Therefore, we get:

Theorem 1.2. *Replacing the Luenberger high-gain observer by the high-gain extended Kalman filter, Theorem 1.1 is still valid, i.e., for any triple of compact subsets* $\Gamma, \Gamma', \Gamma''$, $\Gamma \times \Gamma' \times \Gamma'' \subset R^n \times R^n \times S_n(+)$, *for* θ *large enough, the high-gain extended Kalman filter coupled with the system* Σ *to which the feedback control* $\tilde{\alpha}(z)$ *is applied, is asymptotically stable within* $\Gamma \times \Gamma' \times \Gamma''$ *at* $(0, 0, S_\infty)$.

Exercise 1.2. Give all details for the proof of Theorem 1.2.

With exactly the same type of argument, it is clear that we can also use the continuous-discrete version of the high-gain extended Kalman filter:

Exercise 1.3. State and prove a version of Theorem 1.2 using the continuous-discrete version of the high-gain EKF.

2. The General Case of a Phase-Variable Representation

2.1. Preliminaries

We will now deal with the case of Chapters 4 and 5, where the system has a phase-variable representation of a certain order. This happens generically if $d_y > d_u$ (at least for bounded and sufficiently differentiable controls, but it will be the case here).

The systems of the previous paragraph have a phase-variable representation, as we have noticed already. So one could ask why the considerations in the previous paragraph? The answer is that the procedure for stabilizing (asymptotically) via output information is **much less complicated in that case.** In particular, now, we will have to deal with a certain number of successive derivatives of the inputs, because we have only $U^{N,B}$ output observers. In the previous paragraph, we had a $U^{0,B}$ state observer, hence, these problems did not appear.

We will obtain the result with the high-gain Luenberger output observer. Generalizations to the case of the high-gain extended Kalman filters (either continuous-continuous or continuous-discrete) can be done in the same way as in the proof of Theorem 1.2 above. We will leave these generalizations as exercises.

2.1.1. Rings of C^∞ Functions

Recall that in Chapter 5 we introduced several rings of (germs of) analytic functions: $\mathfrak{R}_N, \hat{\mathfrak{R}}_N, \check{\mathfrak{R}}_N, \check{\mathfrak{R}}$. Recall that $\hat{\mathfrak{R}}_N$ was just the pull back ring:

$\hat{\mathfrak{R}}_N = (S\Phi_N^\Sigma)^*(O_{z_0})$, where $z_0 = S\Phi_N^\Sigma(x_0, u_0^{(0)}, \widetilde{u}_{0N-1})$. We will consider the C^∞ analogs $\tilde{\mathfrak{R}}_N, \bar{\mathfrak{R}}_N, \tilde{\mathfrak{R}}$ of the rings $\hat{\mathfrak{R}}_N, \mathfrak{R}_N, \check{\mathfrak{R}}$, i.e.,

$$\tilde{\mathfrak{R}}_N\left(x_0, u^{(0)}, \tilde{u}_{N-1}\right) = \left\{G \circ S\Phi_N^\Sigma\right\},$$

where G varies over the germs of C^∞ functions at the point $S\Phi_N^\Sigma(x_0, u^{(0)}, \tilde{u}_{N-1})$, and

$$\bar{\mathfrak{R}}_N\left(x_0, u^{(0)}, \tilde{u}_{N-1}\right) = \left\{G \circ S\Phi_{N,\tilde{u}_{N-1}}^\Sigma\right\},$$

where G varies over the germs of C^∞ functions at the point $S\Phi_{N,\tilde{u}_{N-1}}^\Sigma(x_0, u^{(0)})$.

The ring $\tilde{\mathfrak{R}}(x_0, u^{(0)})$, or simply $\tilde{\mathfrak{R}}$, if there is no ambiguity, will be the ring of germs of functions of the form

$$G(u, \varphi_1, \ldots, \varphi_p)$$

at the point $(x_0, u^{(0)})$, where G is C^∞ and all the functions φ_i are of the form

$$\varphi_i = L_{f_u}^{k_1}(\partial_{j_1})^{s_1} L_{f_u}^{k_2}(\partial_{j_2})^{s_2} \ldots\ldots L_{f_u}^{k_r}(\partial_{j_r})^{s_r} h.$$

(Again, $\partial_j = (\frac{\partial}{\partial u_j})$).

Recall that, in the analytic case, the condition $ACP(N)$ is equivalent to $\check{\mathfrak{R}} = \mathfrak{R}_N \subset \hat{\mathfrak{R}}_N$ by Theorem 4.1 of Chapter 5. Of course, if $ACP(N)$ holds, then a fortiori,

$$\tilde{\mathfrak{R}} = \bar{\mathfrak{R}}_N \subset \tilde{\mathfrak{R}}_N : \tag{119}$$

Because the above generators u, φ_i of $\tilde{\mathfrak{R}}$ belong to $\mathfrak{R}_N \subset \hat{\mathfrak{R}}_N$, hence, C^∞ functions of them belong to $\bar{\mathfrak{R}}_N$. Therefore $\tilde{\mathfrak{R}} \subset \bar{\mathfrak{R}}_N$. Using the same reasoning, $\tilde{\mathfrak{R}} \subset \bar{\mathfrak{R}}_N \subset \tilde{\mathfrak{R}}_N$. Also, by definition, $\bar{\mathfrak{R}}_N \subset \tilde{\mathfrak{R}}$.

Moreover, the analyticity assumption plays no role in this result, so that it is also true for a C^∞ system Σ that $ACP(N)$ is equivalent to the condition (119).

2.1.2. Assumptions

We will start with the most general situation, that is, we assume that our system Σ satisfies the assumptions of Theorem 5.2 of Chapter 5, i.e.,

(H_1) Σ satisfies $ACP(N)$ at each point,

(H_2) Σ is differentially observable of order N.

In particular, if Σ is strongly differentially observable of order N (the "generic" situation of Chapter 4), these assumptions are satisfied.

For the same reasons as above, we assume also that $X = R^n$.

In this chapter, Σ is semiglobally asymptotically stabilizable at $(x_0 = 0, u_0 = 0)$. We will have to make an additional assumption (H_3), relative to the stabilizing feedbacks α_Γ:

(H_3) The germs of the $\alpha_{\Gamma,j}(.)$ at x_1, $j = 1, \ldots, d_u$, belong to $\bar{\mathfrak{R}}_N(x_1, u^{(0)}, \tilde{u}_{N-1})$, for all $x_1 \in X = R^n$ and for all $(u^{(0)}, \tilde{u}_{N-1}) \in U \times R^{(N-1)d_u}$. Equivalently, these germs belong to $\bar{\mathfrak{R}}$ by virtue of (119).

2.1.3. Comments

1. This assumption (H_3), (together with (H_1) and (H_2)), **is automatically satisfied if Σ is strongly differentially observable of order** N: in that case, the ring $\bar{\mathfrak{R}}_N$ is just the ring of germs of smooth functions at $(x_1, u^{(0)})$.

Exercise 2.1. Prove this last statement.

2. We have defined the C^∞ analogs $\bar{\mathfrak{R}}_N, \bar{\mathfrak{R}}_N, \bar{\mathfrak{R}}$ of our rings $\hat{\mathfrak{R}}_N, \mathfrak{R}_N, \check{\mathfrak{R}}$ for the following reasons. First, in this section, we want to deal with C^∞ systems (recall that the results of Chapter 5 are valid also in the C^∞ case). Second, it could happen that, **even if Σ is analytic** and asymptotically stabilizable, then it is asymptotically stabilizable by a feedback that is only C^∞.
3. At this point, it is important to say a few words about U. In the previous section, we assumed that $U = I^{d_u}$, where I is a closed interval of R. Assume that I is not equal to R. Then we can find a diffeomorphism $\Psi : Int(I) \rightarrow R$. The rings above are intrinsic objects that do not depend on coordinates on U. So that $ACP(N)$ does not depend on a change of variable over u. On the same way, the assumption (H_2) is intrinsic. Also, the fact that the germ of $\alpha_{\Gamma,j}$ belongs to $\bar{\mathfrak{R}}_N$ at each point is intrinsic. Therefore, because our stabilizing feedbacks α_Γ take their values in the interior of U, we see that **we can replace u by v,** $v_i = \Psi(u_i)$ **and assume that** $I = R$. This is what we will do in the remainder of this section.

2.1.4. A Crucial Lemma

In order to state and prove our main theorem in this section, we need a preliminary result:

Lemma 2.1. *Assume that Σ is given, and that (H_1) and (H_3) are satisfied. Then, (H_1) and (H_3) are also satisfied for Σ^r, the rth dynamical extension of Σ.*

The definition of the rth dynamical extension of Σ is given in Chapter 2, Definition 4.1.

Proof. It is obvious that, if Σ satisfies (H_1), so does Σ^r.

From the definitions of $\bar{\mathfrak{R}}_N$ and of the rth dynamical extension of Σ, we see that it is sufficient to prove the lemma for $r = 1$.

$$(\Sigma^1)\begin{cases} \dot{x} = f(x, u), \\ \dot{u} = u_1, \\ y = (h(x, u), u). \end{cases}$$

A compact $\Gamma \subset X$ being given, we consider $\alpha_\Gamma : X \to R^{d_u}$ satisfying (H_3), that is, the origin is asymptotically stable within Γ for the vector field (112). Let B_Γ be the basin of attraction of the origin for (112).

By the inverse Lyapunov's theorems ([35, 40]), there is a proper C^∞ Lyapunov function for (112), i.e., (i) V is proper over B_Γ, $V(0) = 0$, $V(x) > 0$ if $x \neq 0$ and (ii) on B_Γ and along the trajectories of (112), $\dot{V} < 0$ except for $x = 0$ where $\dot{V}(0) = 0$. We will also consider some arbitrary interval $I_B = [-B, B]$ and prove that we can construct a feedback for Σ^1 with the required properties, and such that the basin of attraction of the origin contains $\Gamma \times (I_B)^{d_u}$.

We choose for Σ^1 the feedback $u_1 = L_f\alpha_\Gamma - r(u - \alpha_\Gamma)$, denoted by α_r, for $r > 0$. Set $\tilde{H}(x, u) = u - \alpha_\Gamma$. We get for the system Σ^1 under feedback:

$$(\Sigma^{1,r})\begin{cases} \frac{dx}{dt} = f(x, \alpha_\Gamma(x) + \tilde{H}), \\ \frac{d\tilde{H}}{dt} = -r\tilde{H}. \end{cases} \tag{120}$$

Consider the function $W(x, \tilde{H}) = V(x) + \frac{1}{2}\sum_{i=1}^{d_u}(\tilde{H}_i)^2$. This function W is proper on $B_\Gamma \times R^{d_u}$. Denote by DW_k the compact set $DW_k = \{(x, u) | W(x, u) \leq k\}$, $k \geq 0$. Let M be the maximum of W on $\Gamma \times (I_B)^{d_u}$. Let $\Gamma' = DW_{M+1}$, $\Gamma'' = DW_M$. Then $\Gamma'' \subset Int(\Gamma')$, Γ'' is compact and contains $\Gamma \times (I_B)^{d_u}$. We set $\Gamma''' = \Gamma' \backslash Int(\Gamma'')$.

Along the trajectories of $\Sigma^{1,r}$, we have

$$\frac{d}{dt}(W(x, \tilde{H})) = dV(x).f(x, \alpha_\Gamma(x) + \tilde{H}) - r\sum_{i=1}^{d_u}(\tilde{H}_i)^2. \tag{121}$$

We claim that if r is sufficiently large, this quantity is strictly negative for $(x, u) \in \Gamma'''$. First, if $V(x)$ is smaller than $M/2$, $\|\tilde{H}\|$ has to be larger than $\sqrt{M}$, and r can be chosen large enough for the quantity (121) above to be negative (on $\Gamma''' \cap \{V(x) < \frac{M}{2}\}$). Second, let us examine the case where $V(x) \geq \frac{M}{2}$:

$$\begin{aligned}\frac{d}{dt}(W(x, \tilde{H})) &= dV(x).f(x, \alpha_\Gamma(x)) \\ &\quad + \sum_{i=1}^{d_u}(u_i - \alpha_{\Gamma,i})dV(x).\Phi_i(x, u - \alpha_\Gamma(x)) \\ &\quad - r\sum_{i=1}^{d_u}(u_i - \alpha_{\Gamma,i})^2 \\ &= -\sum_{i=1}^{d_u}\left[(u_i - \alpha_{\Gamma,i})\sqrt{r} - \frac{1}{2\sqrt{r}}dV(x).\Phi_i(x, u - \alpha_\Gamma(x))\right]^2 \\ &\quad + dV(x).f(x, \alpha_\Gamma(x)) + \frac{1}{4r}\sum_{i=1}^{d_u}(dV(x).\Phi_i)^2.\end{aligned}$$

But, for r large enough, $\frac{1}{4r}\sum_{i=1}^{d_u}(dV(x).\Phi_i)^2$ is smaller than the minimum of $|dV(x).f(x, \alpha_\Gamma(x))|$ over $\Gamma''' \cap \{V(x) \geq \frac{M}{2}\}$. This shows that the semitrajectories of $\Sigma^{1,r}$ starting from Γ''' stay in Γ'.

Now, going back to the expression (120) of $\Sigma^{1,r}$, let us show that it is asymptotically stable at the origin:

$H^0 = \{\tilde{H} = 0\}$ is an invariant manifold for $\Sigma^{1,r}$. On H^0, the dynamics is $\dot{x} = f(x, \alpha_\Gamma(x))$, hence, it is asymptotically stable at $x = 0$ by assumption. If the linearized vector field at $x = 0$ is asymptotically stable, the same holds for $\Sigma^{1,r}$ obviously. If it is not, then we pick a (asymptotically stable) center manifold C of $\dot{x} = f(x, \alpha_\Gamma(x))$. Obviously, $C \times \{0\}$ is also an asymptotically stable center manifold for $\Sigma^{1,r}$. Hence, $\Sigma^{1,r}$ is asymptotically stable at the origin.

Let us now consider a semitrajectory $\{x(t), u(t) | t \geq 0\}$ of $\Sigma^{1,r}$ starting from $\Gamma \times (I_B)^{d_u}$. We know that this semitrajectory stays in Γ', and hence, is bounded. Its ω-limit set Ω is nonempty. Let $(x^*, \alpha_\Gamma(x^*))$ be a point in Ω. (It should be of this form, because $u(t) - \alpha_\Gamma(x(t)) \to 0$). Because Ω is invariant, the dynamics on Ω is given by $\dot{x} = f(x, \alpha_\Gamma(x))$, which is asymptotically stable within B_Γ by assumption. Hence, the trajectory issued from $(x^*, \alpha_\Gamma(x^*))$ is in the basin of attraction of the origin of $\Sigma^{1,r}$. Since the basin of attraction is open, the same is true for the trajectory $(x(t), u(t))$. It follows that $(x^*, \alpha_\Gamma(x^*)) = 0$.

We have proven that Σ^1 is asymptotically stabilizable at the origin within $\Gamma \times I_B$, with the feedback $u_1 = \alpha_r(x, u)$, for r large enough. Recall that $\alpha_r(x, u) = L_f\alpha_\Gamma - r(u - \alpha_\Gamma)$. By our assumptions, $\alpha_{\Gamma,j} \in \bar{\mathfrak{R}}_N = \bar{\mathfrak{R}}$ (relative to Σ). By definition, $\bar{\mathfrak{R}}$ is stable by L_f. Hence, the germs of $\alpha_{r,j}(x, u) \in \bar{\mathfrak{R}}_N(\Sigma)$.

Now we will show that: (i) α_r is a function of the state of Σ^1 only, and (ii) α_r belongs to $\tilde{\Re}_N(\Sigma^1)$. This implies that α_r is in $\bar{\Re}_N(\Sigma^1)$, because an element of $\tilde{\Re}_N(\Sigma^1)$, which does not depend on the derivatives of the control of Σ^1, is in $\bar{\Re}_N(\Sigma^1)$ by definition. Hence, $\alpha_{r,j}(x,u) \in \bar{\Re}_N(\Sigma^1)$. Number (i) is obvious; and (ii) because Σ satisfies the $ACP(N)$, we know by (119) that $\bar{\Re}_N(\Sigma) \subset \tilde{\Re}_N(\Sigma)$. Also, $\tilde{\Re}_N(\Sigma) \subset \tilde{\Re}_N(\Sigma^1)$, and $\alpha_{r,j}(x,u) \in \bar{\Re}_N(\Sigma)$, hence, $\alpha_{r,j}(x,u) \in \tilde{\Re}_N(\Sigma^1)$. This completes the proof. ■

Corollary 2.2. *If Σ satisfies (H_1), (H_3), then Σ^N is stabilizable within any arbitrary compact set Γ with a feedback α_Γ that belongs to $\tilde{\Re}_N(\Sigma)$ (the germs of which belong to $\bar{\Re}_N(\Sigma)$).*

Proof. Obvious. ■

2.2. Output Stabilization Again

First, we will consider the Luenberger version of the output observer. We set $M = N(d_y + d_u)$, where N is such that Σ satisfies the condition $ACP(N)$ and is differentially observable of order N. $R^M = R^{N(d_y+d_u)}$.

We choose $\Gamma \subset X$, $\Gamma' \subset R^{Nd_y}$, $\Gamma'' \subset R^{Nd_u}$, three arbitrary compact sets. We denote (as in the previous chapters) by $A_{m,p}$ the (mp, p) block-antishift matrix, i.e., $A_{m,p} : R^{mp} \to R^{mp}$,

$$A_{m,p} = \begin{pmatrix} 0, Id_p, 0, \ldots\ldots\ldots, 0 \\ \cdot \\ \cdot \\ 0, \ldots\ldots\ldots\ldots, 0, Id_p \\ 0, \ldots\ldots\ldots\ldots\ldots, 0 \end{pmatrix}.$$

Also, $C_{m,p} : R^{mp} \to R^p$ and $b_{m,p} : R^p \to R^{mp}$ denote the matrices,

$$C_{m,p} = (Id_p, 0, \ldots\ldots, 0),$$

$$b_{m,p} = \begin{pmatrix} 0 \\ \cdot \\ \cdot \\ \cdot \\ 0 \\ Id_p \end{pmatrix}.$$

We take a feedback $\alpha^N_{\Gamma\times\Gamma''}$, given by Corollary 2.2, that stabilizes the Nth dynamical extension Σ^N of Σ within $\Gamma \times \Gamma''$. Recall that Σ^N is given by

$$\begin{aligned} \dot{x} &= f(x,u), \\ \dot{\omega} &= A_{N,d_u}\omega + b_{N,d_u}u_N, \end{aligned} \tag{122}$$

with the notation $\omega = (u^{(0)}, \tilde{u}_{N-1})$.

Then, the feedback system

$$\begin{aligned}\dot{x} &= f\big(x, u^{(0)}\big),\\ \dot{\omega} &= A_{N,d_u}\omega + b_{N,d_u}\alpha^N_{\Gamma\times\Gamma''}(x,\omega),\end{aligned} \tag{123}$$

is asymptotically stable at $(x_0, 0)$ within $\Gamma \times \Gamma''$. Let $\tilde{B}$ denote the basin of attraction of $(x_0, 0)$ for (123).

Let V be a proper Lyapunov function for this vector field on $\tilde{B}$. The function V has a maximum m over $\Gamma \times \Gamma''$. Setting $D_k = \{s | V(s) \leq k\}$, $k \geq 0$, let us consider D_m, D_{m+1}. Then, $\Gamma \times \Gamma'' \subset D_m \subset Int(D_{m+1})$.

Using the C^∞ version of Corollary 5.3 of Chapter 5, we can find a C^∞ function α defined on all of R^M such that

$$\alpha^N_{\Gamma\times\Gamma''}\big(x, u^{(0)}, \tilde{u}_{N-1}\big) = \alpha\big(S\Phi^\Sigma_N\big(x, u^{(0)}, \tilde{u}_{N-1}\big)\big), \tag{124}$$

for all $(x, u^{(0)}, \tilde{u}_{N-1}) \in D_{m+1}$, and moreover, α can be taken compactly supported.

Hence, α reaches its maximum over R^M. So do $u^{(0)}$, $u^{(i)}$, $1 \leq i \leq N-1$ over D_{m+1}. Let B be the maximum of these maxima. Let $\tilde{\Gamma}$ be the image of D_{m+1} by the projection $\Pi_1 : X \times R^{Nd_u} \to X$. The set $\tilde{\Gamma}$ is compact.

We consider the $\varphi^{\tilde{\Gamma}}_N$ given by Theorem 5.2 of Chapter 5, applied to Σ and $\tilde{\Gamma}$, i.e.,

$$y^{(N)} = \varphi^{\tilde{\Gamma}}_N\big(y^{(0)}, \tilde{y}_{N-1}, u^{(0)}, \tilde{u}_N\big),$$

for all $x \in \tilde{\Gamma}$, all $u^{(0)}$, $\tilde{u}_N$.

We can now "**couple**" the feedback $\alpha^N_{\Gamma\times\Gamma''}$ with the $U^{N,B}$ output observer in its Luenberger form (Formula (81), Section 2.3, Chapter 6).

Let us write the equation of the full **coupled** system over $X \times R^M$:

$$\begin{aligned}&(i)\ \dot{x} = f\big(x, u^{(0)}\big),\\ &(ii)\ \dot{\omega} = A_{N,d_u}\omega + b_{N,d_u}\alpha(z,\omega),\\ &(iii)\ \dot{z} = (A_{N,d_y} - K_\theta C_{N,d_y})z + K_\theta h\big(x, u^{(0)}\big) + b_{N,d_y}\varphi^{\tilde{\Gamma}}_N(z, \omega, \alpha(z,\omega)).\end{aligned} \tag{125}$$

Our result will be the following, as expected:

Theorem 2.3. *Assumptions* (H_1), (H_2), *and* (H_3) *are made. For any* $\Gamma' \subset R^{Nd_y}$, *for* θ *large enough, the system (125) is asymptotically stable within* $\Gamma \times \Gamma'' \times \Gamma'$ *at* $(x_0, 0, 0)$.

This theorem means that in all the cases we have dealt with in the previous chapters, we can stabilize asymptotically, using output information only,

within arbitrarily large compact sets as soon as we can stabilize asymptotically within compact sets by smooth-state feedback.

In particular, if the system Σ is strongly differentially observable of some order N, which is generic if $d_y > d_u$, and if Σ is smooth state feedback stabilizable, this theorem applies.

The proof of this theorem will go through the three same steps as the proof of Theorem 1.1.

Proof of Theorem 2.3. Let us observe first that in $X \times R^M$, the set $S = \{z = \Phi_N^{\Sigma}(x, \omega)\}$ is an embedded submanifold (it is the graph of a smooth mapping).

Set $S' = (D_{m+1} \times R^{Nd_y}) \cap S$. It is clear that S' is an invariant manifold for (125): If we set $Y = \Phi_N^{\Sigma}(x, \omega)$, and $\varepsilon = z - Y$, remember that the equation of ε over D_{m+1} is given by Equations (84) or (85) of Chapter 6. If we rewrite it using the notations of this chapter, it gives:

$$\frac{d\varepsilon}{dt} = (A_{N,d_y} - K_\theta C_{Nd_y})\varepsilon + b_{N,d_y}\big(\varphi_N^{\tilde{\Gamma}}(z, \omega, \alpha(z, \omega)) - b_{N,d_y}\varphi_N^{\tilde{\Gamma}}(Y, \omega, \alpha(z, \omega))\big), \tag{126}$$

so that, if $\varepsilon = 0$, it is just $\frac{d\varepsilon}{dt} = 0$.

Moreover, by construction, $Int((D_{m+1} \times R^{Nd_y}) \cap S) = \{(x, \omega, z) | (x, \omega) \in Int(D_{m+1}), z = \Phi_N^{\Sigma}(x, \omega)\}$. If $\varepsilon = 0$, D_{m+1} is positively invariant by the dynamics (i) and (ii) in (125), because it is the dynamics of an asymptotically stable system on $X \times R^{Nd_u}$, and D_{m+1} is a sublevel set of a Lyapunov function. It follows that in fact $S' = (D_{m+1} \times R^{Nd_y}) \cap S$ **is a compact manifold with a boundary, which is positively invariant under the dynamics (125).**

To summarize, setting $S'' = Int(S')$, S'' is an embedded manifold with the same property that it is positively invariant under (125). The point $(x_0, 0, 0) \in S''$ and the dynamics on S'' (and on S') is globally asymptotically stable.

2.2.1. *First Step: Local Asymptotic Stability of (125)*

The linearized dynamics for ε at $(x_0, 0, 0)$ is, after straightforward computations,

$$\dot{\varepsilon} = (A_{N,d_y} - K_\theta C_{N,d_y})\varepsilon + b_{N,d_y} L\varepsilon. \tag{127}$$

Exercise 2.2. Using the same time reparametrization method as in the proof of Theorem 2.5 of Chapter 6, prove that for θ large enough, this linear vector field is asymptotically stable.

Again, if the linearized dynamics on S'' is asymptotically stable, the same is true for (125): All the eigenvalues of the linearized vector field at the equilibrium have negative real part. If the linearized dynamics on S'' is not asymptotically stable, again, the asymptotically stable center manifold for the dynamics restricted to S'' is also a center manifold for the full system (125). It follows that (125) is asymptotically stable at $(x_0, 0, 0)$.

2.2.2. Second Step: All the Semitrajectories Staying in $D_{m+1} \times R^{Nd_y}$ Are Contained in the Basin of Attraction of $(x_0, 0, 0)$

Let $\{p(t) = (x(t), \omega(t), z(t)),\ t \geq 0\}$ be such a semitrajectory. Theorem 2.5 of Chapter 6 applies: $|\omega_{i,j}| \leq B$ for $1 \leq i \leq N,\ 1 \leq j \leq d_u$, Hence, $\lim_{t\to+\infty} \varepsilon(t) = 0$. Therefore, $\{z(t),\ t \geq 0\}$ is bounded. So that the ω-limit set Ω of $\{p(t), t \geq 0\}$ is nonempty. Let $p^* = (x^*, \omega^*, z^*)$ be in Ω. Because $\varepsilon(t) \to 0$, it follows that $z^* = \Phi_N^{\Sigma}(x^*, \omega^*)$. Hence, $p^* \in S'$. S' is positively invariant, and every semitrajectory starting in S' tends to zero, as we said. Thus, the semitrajectory starting from p^* is in the basin of attraction of zero. The same happens to the semitrajectory $p(t)$, and at the end, $\lim_{t\to+\infty} p(t) = (x_0, 0, 0)$.

2.2.3. Third Step: All Semitrajectories Starting From $\Gamma \times \Gamma'' \times \Gamma'$ Stay in $D_{m+1} \times R^{Nd_y}$

The argument is exactly the same as in the proof of Theorem 1.1. These trajectories take a certain time $T_{\min}$ to cross $(D_{m+1} \times R^{Nd_y})\backslash(D_m \times R^{Nd_y})$: in this set, $u^{(N)}(t)$ is bounded by B. This time can be used if we increase θ, to make ε very small at time $T_{\min}$ and afterward. (Here we use again "the exponential against the polynomial," in the estimation of the error in the output observer. We use also the fact that because $z(0)$, $\omega(0)$, and $x(0)$ belong to given compact sets, the error at time 0 has a bound.) Now, we consider our Lyapunov function V over the boundary ∂D_{m+1}. Along our trajectories, when $\varepsilon = 0$, $\frac{d}{dt} V(x(t), \omega(t))$ is strictly smaller than a certain negative constant $-\zeta^2$ (because ∂D_{m+1} is compact and does not contain $(x_0, 0)$).

However, when $\varepsilon(t)$ *is* nonzero, Equations (125) (i) and (ii) can be rewritten as

$$(i)\ \dot{x} = f\big(x, u^{(0)}\big),$$
$$(ii)\ \dot{\omega} = A_{N,d_u}\omega + b_{N,d_u}\alpha\big(\Phi_N^{\Sigma}(x(t), \omega(t)) + \varepsilon(t), \omega(t)\big).$$

Hence, if $\varepsilon(t)$ is sufficiently small, $\frac{d}{dt} V(x(t), \omega(t))$ is still strictly negative. This just means that $(x(t), \omega(t))$ stays in D_{m+1}.

By virtue of step 2 above, all these semitrajectories tend to $(x_0, 0, 0)$. By virtue of step 1, Equations (125) are asymptotically stable at $(x_0, 0, 0)$, and the basin of attraction contains $\Gamma \times \Gamma'' \times \Gamma'$. ■

Exercise 2.3. Consider the system Σ of Exercise 4.1 in Chapter 5:

$$X = R^2, U = R, y = h(x) = x_1,$$
$$\dot{x}_1 = x_2^3 - x_1,$$
$$\dot{x}_2 = x_2^8 + x_2^4 u.$$

1. Show that the feedback $u_r(x) = -r(x_2)^3$, $r > 0$, stabilizes asymptotically Σ at $(0, 0)$. Show that the basin of attraction is $b_r = \{x | x_2 < r\}$.
2. Show that the previous theorem applies, but Σ **is not strongly differentially observable, of any order.**

Theorem 2.4. *The statement of Theorem 2.3 also holds when we replace the Luenberger version of the output observer by its extended Kalman filter version (Corollary 2.9 of Chapter 6).*

Exercise 2.4. Give a proof of Theorem 2.4.

Exercise 2.5. State and prove a version of Theorem 2.4, using the high-gain extended Kalman filter in its continuous-discrete form (Theorem 2.20 of Chapter 6).

3. Complements

3.1. Systems with Positively Invariant Compact State Spaces

This is a situation that seems to be very common in practice. In particular, it will appear in the first application of the next chapter. We make the additional assumption:

(H_4): (i) The system Σ is such that the state space is not $X = R^n$, but a certain relatively compact open subset $\Omega \subset R^n$. We assume that Σ is also defined and smooth on the boundary $\partial\Omega$, and the closure $Cl(\Omega)$ is positively invariant for the dynamics of Σ whatever the control $u(.)$, with values in U.

(ii) The state feedback is asymptotically stabilizing within $Cl(\Omega)$.

Observation. (H_4, (i)) implies that Ω also is positively invariant for the dynamics of Σ.

Proposition 3.1. *In the case in which the assumption (H_4) holds, all the theorems in the two previous sections in this chapter are true with a refinement: As soon as the observers are exponential,* ***with any rate of decay of the error,*** *the full system controller+observer is asymptotically stable within $Cl(\Omega) \times \hat{X}$, where $\hat{X}$ is the state space of the observer.*

Proof.

1. The proof of local asymptotic stability is the same.
2. The proof of the fact that bounded semitrajectories tend to zero is the same.
3. The semitrajectories of the full coupled system are automatically bounded: (a) $x(t)$, the state of Σ, is bounded, by assumption; (b) the observer error going to zero, the estimate is also bounded; (c) (in the case of the high-gain EKF only), the component $S(t)$ is bounded and tends to S^{∞} exactly for the same reason as in the general case. ■

Exercise 3.1. Give details of the previous proof, especially in the case of the high-gain extended Kalman filter.

This is very important in practice, because, as we explained, the observer is not **doubly high-gain** in that case.

8

Applications

In this chapter, to illustrate the theoretical content of this book we will briefly present and comment on two applications of our theory to the real world of chemical engineering. The first application was really applied in practice. The second was done in simulation only.

1. Binary Distillation Columns

Distillation columns are generic objects in the petroleum industry, and more generally, in chemical engineering. Here we limit ourselves to the case of binary distillation columns, i.e., distillation columns treating a mixture of two components. The example we consider in practice is a "depropanizer" column, i.e., a column distillating a mixture of butane and propane. There is at least one of these columns in each petroleum refinery. This work has been also generalized to some cases of multicomponent (nonbinary) distillation, but this topic will not be discussed here. For more details on this application, the reader should consult [64].

1.1. Presentation of a Distillation Column, the Problems, and the Equations

1.1.1. Binary Distillation Columns

The column is represented in Figure 1. It is composed of a certain number of trays (n), a top condenser and a boiler. The top condenser is considered as the tray 1, and the bottom of the column is considered as the tray n. In practice, n is approximately 20.

The feed (here, the butane-propane mixture) enters the column at a medium tray (the fth one, $f \cong 10$). The light component (propane) is expected to be extracted at the top of the column (distillate). The heavy component (butane) is the bottom product (residue) on the figure. The distillate and the residue

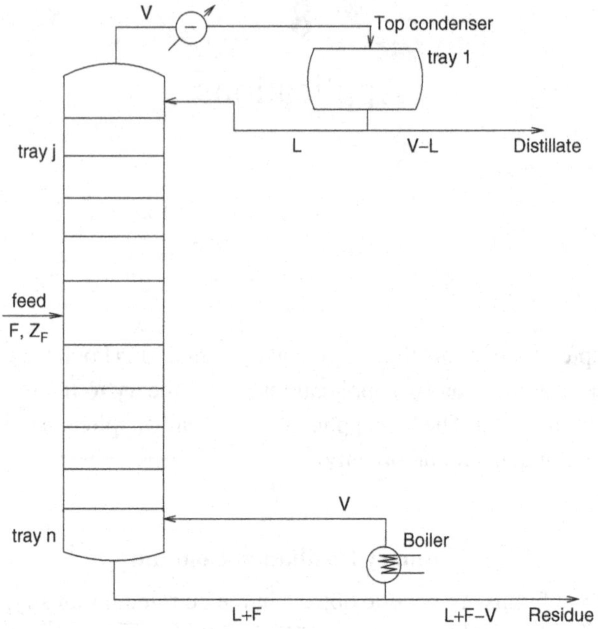

Figure 1. Schematic of a distillation column.

are partly recycled in the column. During this recycling, the residue is boiled, and the distillate is condensed.

On each tray j arrive the vapor coming from tray $j + 1$ and the liquid coming from tray $j - 1$. A certain (constant) amount H_j of liquid mixture is on the tray j at time t.

The assumptions that are made are the standard **Lewis hypotheses** (see [54] for justification). Under these assumptions, the flowrate of liquid going down in the column from the top condenser is constant (doesn't depend on the tray j) above the feed and below the feed: it is L above the feed, and $L + F$ below the feed (the feed mixture is assumed to be liquid and enter the column at its "bubble point"). The flowrate of vapor is constant (does not depend on the tray j) and equal to V all along the column.

The control variables are $L(t)$ and $V(t)$, the "reflux flowrate" and "vapor flowrate" ($L(t)$ is controlled directly by the user, and $V(t)$ is controlled via the boiler at the bottom of the column).

On each tray j:

- In the liquid phase, the composition (molar fraction) of the light component (propane) is denoted by x_j, the composition of the heavy component (butane) is $1 - x_j$.

- In the vapor phase, the composition of the light component on tray j is denoted by y_j, and the composition of the heavy component is just $1 - y_j$.

On each tray, liquid-vapor equilibrium is assumed, and y_j is related to x_j by a certain function k:

$$y_j = k(x_j). \tag{128}$$

The function k is called the **liquid-vapor equilibrium law**. It comes from thermodynamics and depends only on the binary mixture (and the pressure in the column, **which is assumed to be constant**).

We assume that the function k is monotonic, $k(0) = 0,\ k(1) = 1$. (This corresponds to what is called nonazeotropic distillation, the most common situation in practice.) Typically, the function k has the following expression:

$$k(x) = \frac{\alpha x}{1 + (\alpha - 1)x}, \tag{129}$$

where $\alpha > 1$ is a constant, depending on the mixture only (and of the pressure), called the "relative volatility" of the mixture.

Now anyone is able to write a material balance on each tray to get the equations for the distillation column.

1.1.2. The Equations of the Column

The **condenser** equation is

$$H_1 \frac{dx_1}{dt} = V(y_2 - x_1). \tag{130}$$

The equation of a general tray above the feed tray is

$$H_j \frac{dx_j}{dt} = L(x_{j-1} - x_j) + V(y_{j+1} - y_j). \tag{131}$$

The equation of the feed tray is

$$H_f \frac{dx_f}{dt} = F(Z_F - x_f) + \ L(x_{f-1} - x_f) + V(y_{f+1} - y_f). \tag{132}$$

The equation of a general tray below the feed is

$$H_j \frac{dx_j}{dt} = (F + L)(x_{j-1} - x_j) + V(y_{j+1} - y_j). \tag{133}$$

The equation of the bottom tray is

$$H_n \frac{dx_n}{dt} = (F + L)(x_{n-1} - x_n) + V(x_n - y_n). \tag{134}$$

This set of equations – (130), (131), (132), (133), and (134) – in which $y_j = k(x_j)$, constitutes the mathematical model of the binary distillation column.

In these equation, x_j, the jth state variable, is the molar fraction of the light product (propane) on tray j. The constant H_j is the molar quantity of liquid phase contained on the tray j (a geometric constant). The variable $Z_F(t)$ is the feed composition (again, molar fraction of the light component); F is the feed flowrate (a constant); $L(t)$, **the first control variable** is the reflux flowrate; and $V(t)$, **the second control variable** is the vapor flowrate in the column.

The column is expected to separate the products of the mixture so that the output variables are x_1, $1 - x_n$ (the quality of the light product at the top of the column, and the quality of the heavy product at the bottom of the column). The output x_1 has to be close to 1, x_n has to be close to zero. Typical target values in the case of a butane–propane mixture are $x_1 = 0.996$, $x_n = 0.004$.

Remark 1.1. This model is very simple. In fact, more sophisticated models have basically the same structure. They include a set of (algebraic) equations, which expresses the **thermal balance** on each plate. This set of equations is replaced here by the **Lewis hypotheses.**

Also, the expressions (128) and (129), are replaced by a more complicated relation describing liquid–vapor equilibria. However, the structure remains the same, and **all the developments presented in this chapter can be extended to these more sophisticated models.**

1.1.3. The Problems to Be Solved

Continuous quality measurements are extremely expensive (taking into account the maintenance of the devices, it costs about $100,000 a year for a single continuous measurement). Hence, it is completely unreasonable to expect to have a continuous measurement of the quality of a product, which is not a final product (to be sold on the market). In particular, in the case of a butane–propane distillation, there is no hope of measuring Z_F. (**This means that we don't know what is entering the column.**) If any quality is measured (continuously), it will be the quality of the output final products, x_1, x_n.

Here, we assume that the functions $x_1(t)$, $x_n(t)$ are **measured continuously**. This is an expensive assumption, but it was almost realized in our case. In fact, we had discrete measurements, each of four minutes. This sampling time has to be compared to the **response time** of about one hour.

First Problem. (a) Using the output observations only, estimate the state $(x_1, \ldots, x_n)(t)$ of the distillation column, (i.e., estimate all compositions on each plate).

(b) Estimate the quality of the feed $Z_F(t)$.

This first problem is purely a problem of state estimation (there is no control at all).

The second problem concerns control: As we shall see (Section 1.2.2), given reasonable target values of the outputs, (x_1^*, x_n^*), there is a **unique** value of the controls (L^*, V^*) corresponding to an equilibrium x^* for these target values (provided that the unknown Z_F is constant).

Second problem. Stabilize asymptotically the column at the equilibrium x^* corresponding to the given target values (x_1^*, x_n^*) of the outputs. If possible, reduce the **response time**. Of course, we have to solve this second problem **using output informations only**.

Notice that in our case, as we shall show, the equilibria are globally asymptotically stable in open loop (i.e., if we fix the control values to (L^*, V^*), and if the feed composition Z_F-which is unknown-, is a constant). That is, adding the feed composition as a state variable satisfying the additional equation $\dot{Z}_F = 0$, we have a trivial stabilizing smooth state feedback,

$$u_1(x, Z_F) = (L^*(Z_F), V^*(Z_F)), \tag{135}$$

However, of course, this feedback is not very good from the point of view of "response time".

Here, we will give solutions to both problems using our theory. These solutions were tested in practice. They perform very well and are robust. In particular, the users were really interested by the observation procedure. Especially, they were interested in the estimation of the quality of the feed of the distillation column.

It is even a bit surprising that very reasonable dynamic state estimations are obtained on the basis of two measurements only, to estimate $\simeq 20$ state variables, even with the Luenberger version of our high gain observers.

In the case of the other application (the polymerization reactor of the next section), we will be more or less in the same situation, moreover the reactor is open loop unstable.

1.2. The Main Properties of the Distillation Column

1.2.1. Invariant Domains

For an obvious physical reason at the bottom of the column, let us assume that $F \geq V - L$. Consider the domain:

$$U_{\phi,\lambda} = \{u = (L, V) \mid 0 < \lambda \leq L \leq L_{\max}, 0 < \phi \leq V - L \\ \leq (V - L)_{\max} \leq F\}. \tag{136}$$

In practice, we will assume that the set of values of control is $U_{\phi,\lambda}$, where ϕ, λ are arbitrarily small but strictly positive constants. This means in particular that the rates L, V, F in the column are all strictly positive. (This assumption prevents the system to be **close to unobservability**: as we shall see, unobservability holds if $F = V = L = 0$).

F and Z_F are assumed to be constant, and $Z_F \in [z_f^-, z_f^+]$, $z_f^- > 0$, $z_f^+ < 1$. In practice, Z_F is not a constant, but its variation during the stabilization process can be neglected. The reason for this, in the case of the butane–propane column, is that the feed mixture is the top output (the distillate) of another much bigger distillation column.

If $a = (a_1, \ldots, a_n)$, $b = (b_1, \ldots, b_n)$, a_i and b_i are small positive real numbers, let us denote by $\Omega_{a,b}$ the cube $\Omega_{a,b} = [a_1, 1 - b_1] \times \ldots \times [a_n, 1 - b_n]$, $\Omega_0 = \Omega_{0,0} = [0, 1]^n$. The cube Ω_0 is the **physical** state space.

Theorem 1.1. *(i) Ω_0 is positively invariant under the dynamics of the distillation column. Moreover, (ii) for any compact $K \subset Int(\Omega_0)$, it does exist a and b such that $K \subset \Omega_{a,b}$ and $\Omega_{a,b}$ is positively invariant.*

Proof. (i) is a consequence of (ii). Let us prove (ii) constructively. For this, we will exhibit (a, b), arbitrarily small, such that $\Omega_{a,b}$ is positively invariant.

We define a as follows: for $j \leq f$, $a_j = j\bar{a}$, $\bar{a} > 0$ small and $a_n = \varepsilon$, $\varepsilon > 0$ small. For $f < j < n$, $a_j = a_{j+1} + q_{j+1}$, $q_j = q_{j+1}k'_M$, $q_n = q$, $q > 0$ small, $k'_M = \sup_{x \in [0,1]} \frac{dk}{dx}$. In the case where the relative volatility is assumed to be constant, i.e., if $k(x)$ has the expression (129), then, $k'_M = \alpha$. We assume that $\frac{\varepsilon}{q}, \frac{q}{\bar{a}}, \frac{\varepsilon}{\bar{a}}$ are small enough.

The vector b is defined in the following way: if $1 \leq j \leq f$, $q_1 = \bar{b}(1 - \varepsilon')$, for $\varepsilon', \bar{b} > 0$ small, $b_1 = \varepsilon'\bar{b}$, $b_2 = \bar{b}$, $b_{j+1} = b_j + q_j$, $q_j = \beta^j q_1$, $\beta > \sup \frac{L}{Vk'(x)}$. If $j > f$, $b_n = c$, $c > 0$ small, $b_{j+1} = b_j + q_j$, $q_{n-1} = \varepsilon' c$, $q_j = \beta' q_{j+1} = (\beta')^{n-j-1} q_{n-1}$, $\beta' < \inf \frac{Vk'(x)}{F+L}$. We assume also that $c/\bar{b}$ is small.

The proof consists of showing that, if $x = (x_1, \ldots, x_n) \in \Omega_0$,

1. if $x_j = a_j$, then, $\frac{dx_j}{dt} > 0$,
2. if $x_j = 1 - b_j$, then, $\frac{dx_j}{dt} < 0$.

This shows that any trajectory starting from the boundary of $\Omega_{a,b}$ enters immediately in the interior of $\Omega_{a,b}$, and proves the result.

Exercise 1.1. Prove 2- above.

Let us prove 1 only:

$j = 1$: from Equation (130),

$$H_1 \left.\frac{dx_1}{dt}\right|_{x_1 = a_1 = \bar{a}} = V(y_2 - \bar{a}) \geq V(k(a_2) - \bar{a})$$

because k is a strictly increasing function and $y_2 = k(x_2)$, $x_2 \geq a_2$. Then, $H_1 \frac{dx_1}{dt}\big|_{x_1 = a_1 = \bar{a}} \geq V(k(2\bar{a}) - \bar{a}) > 0$.

$2 \leq j \leq f - 1$: the equation of x_j is the Equation (131). Hence,

$$\begin{aligned} H_j \left.\frac{dx_j}{dt}\right|_{x_j = a_j = j\bar{a}} &= L(x_{j-1} - j\bar{a}) - V(k(j\bar{a}) - k(x_{j+1})), \\ &\geq L((j-1)\bar{a} - j\bar{a}) - V(k(j\bar{a}) - k((j+1)\bar{a})) \\ &\geq -L\bar{a} + V\bar{a}k'(x_a), \end{aligned}$$

for some x_a close to 0. Hence, $k'(x_a) > 1$. Because $V - L > 0$, $H_j \frac{dx_j}{dt}\big|_{x_j = a_j = j\bar{a}} > 0$.

$j = f$: by Equation (132),

$$\begin{aligned} H_f \left.\frac{dx_f}{dt}\right|_{x_f = f\bar{a}} &= F(Z_F - f\bar{a}) + L(x_{f-1} - f\bar{a}) + V(k(x_{f+1}) - k(f\bar{a})) \\ &\geq F(Z_F - f\bar{a}) + L((f-1)\bar{a} - f\bar{a}) + V(k(a_{f+1}) - k(f\bar{a})) \\ &\geq FZ_F - Ff\bar{a} - L\bar{a} - Vk'(x_a)\left(f\bar{a} - \varepsilon - q\frac{1-\alpha^{n-f-1}}{1-\alpha}\right) \end{aligned}$$

for some x_a close to zero. Because $\bar{a}, q, \varepsilon$ are small, this quantity is strictly positive.

$j = f + 1$: Equation (133) for $j = f + 1$ gives

$$\begin{aligned} H_{f+1} \left.\frac{dx_{f+1}}{dt}\right|_{x_{f+1} = a_{f+1}} &= (F + L)(x_f - a_{f+1}) + V(k(x_{f+2}) - k(a_{f+1})) \\ &\geq (F + L)(a_f - a_{f+1}) + V(k(a_{f+2}) - k(a_{f+1})) \\ &\geq (F + L)\left(f\bar{a} - \varepsilon - q\frac{1-\alpha^{n-f-1}}{1-\alpha}\right) \\ &\quad - Vk'(x_a)\alpha^{n-f-2}q = B. \end{aligned}$$

However, $B > 0$ is equivalent to

$$1 > \frac{Vk'(x_a)}{(F+L)f}\alpha^{n-f-2}\frac{q}{\bar{a}} + \frac{1}{f}\left(\frac{\varepsilon}{\bar{a}} + \frac{q}{\bar{a}}\frac{1-\alpha^{n-f-1}}{1-\alpha}\right),$$

which is true because $\frac{\varepsilon}{\bar{a}}$ and $\frac{q}{\bar{a}}$ are both small.

$f+2 \le j \le n-1$: by Equation (133),

$$\begin{aligned} H_j \left.\frac{dx_j}{dt}\right|_{x_j=a_j} &= (F+L)(x_{j-1}-a_j) + V(k(x_{j+1}) - k(a_j)) \\ &\ge (F+L)(a_{j-1}-a_j) + V(k(a_{j+1}) - k(a_j)) \\ &\ge (F+L)q_j - Vk'(x_a)q_{j+1} \ge ((F+L)\alpha - Vk'(x_a))q_{j+1}, \end{aligned}$$

for some x_a close to zero. $\alpha > k'(x_a)$ and $F+L > V$, hence, this quantity is strictly positive.

$j = n$: Equation (134) gives

$$\begin{aligned} H_n \left.\frac{dx_n}{dt}\right|_{x_n=a_n=\varepsilon} &= (F+L)(x_{n-1}-\varepsilon) + V(\varepsilon - k(\varepsilon)) \\ &\ge (F+L)(a_{n-1}-\varepsilon) + V(\varepsilon - k(\varepsilon)) \\ &\ge (F+L)q + V(\varepsilon - \varepsilon k'(x_a)), \end{aligned}$$

for x_a close to zero. The ratio $\frac{\varepsilon}{q}$ is small, hence, this quantity is > 0. ■

1.2.2. Stationary Points

We assume that x_1, x_n, F, Z_F are given. Let us set: $\beta = \frac{Z_F - x_n}{x_1 - x_n}$. Typically, at target stationary points, x_1 is close to 1, x_n is close to zero, and Z_F is medium: the "quality" Z_F of the feed is close to .5 (between 0.2 and 0.8, say). Then, β is "medium" (between 0.2 and 0.8).

Let us consider the following open domain D in R^2 (Z_F being fixed):

$$D = \{(x_1, x_n); k^{\circ(n-1)}(x_n) > x_1, k(x_n(1-\beta)+\beta) < x_1, x_n < Z_F < x_1\},$$

where

$$k^{\circ(n-1)}(x_n) = k \circ k \circ \ldots . \circ k(x_n)\,(n-1 \text{ times}).$$

The second and third conditions are always satisfied for target stationary points: $x_n(1-\beta)+\beta$ is close to β, and $k(\beta) < 1$.

The equilibrium coefficient $k(x) = \frac{\alpha x}{1+(\alpha-1)x}$, and α is between 1 and 3 for any binary mixture "not easy to distillate," in particular for the butane–propane mixture.

As we shall see, the first condition in the definition of D is a necessary condition for (x_1, x_n) correspond to an equilibrium. It is necessary and sufficient if the second condition is met.

First, the sum of all (stationary) equations gives the global balance of light product over the column: $V = \beta F + L$. By considering the equations, 1 to $f - 1$, we can compute $x_j,\ 2 \leq j \leq f$:

$$\begin{aligned}&(i)\ y_2 = x_1,\\ &(ii)\ y_j = x_1\left(1 - \frac{L}{L+\beta F}\right) + \frac{L}{L+\beta F}x_{j-1},\end{aligned} \tag{137}$$

and using the equations of the trays n to $f + 1$, we compute $x_j,\ f \leq j \leq n$:

$$\begin{aligned}&(i)\ x_{n-1} = x_n\left(1 - \frac{\beta F + L}{F + L}\right) + \frac{\beta F + L}{F + L}k(x_n),\\ &(ii)\ x_{n-j} = x_n\left(1 - \frac{\beta F + L}{F + L}\right) + \frac{\beta F + L}{F + L}k(x_{n-j+1}),\end{aligned} \tag{138}$$

Exercise 1.2. Prove Formulas (137) and (138).

Exercise 1.3. Prove that the stationary points satisfy $x_1 > x_2 > \ldots. > x_n$.

Using these formulas, it is easy to compute two different estimations x_{f1} and x_{fn} of x_f, and to show the following proposition.

Proposition 1.2. *(i) $x_{f1}(L)$ is a smooth function of L, with strictly negative derivative, and:*

$$x_{f1} :]0, +\infty[\to]k^{\circ(-f+1)}(x1), k^{-1}(x_1)[,$$

(ii) $x_{f2}(L)$ is a smooth function of L, with strictly positive derivative and:

$$x_{f2} :]0, +\infty[\to]\tilde{x}, k^{\circ(n-f)}(x_n)[, \text{ where } \tilde{x} < (1-\beta)x_n + \beta.$$

Exercise 1.4. Prove Proposition 1.2.

If (x_1, x_n) belong to D, then, these two intervals overlap. It follows that the following proposition holds:

Proposition 1.3. *There is a smooth mapping: $u^s_{Z_F} : D \to (R^+)^2$, which to (x_1, x_n) associates $(L, V) = u^s_{Z_F}(x_1, x_n)$, the value of stationary controls* *corresponding to the target stationary outputs* (x_1, x_n). *The mapping* $u^s_{Z_F}$ *is in fact smooth w.r.t.* (z_F, x_1, x_n).

Now conversely, let us examine **what are the equilibria for** L, V **fixed.** Summing the (stationary) equations of the column, we have again, using $V - L > 0$ and $F + L - V > 0$:

$$0 < \frac{V - L}{F} = \beta = \frac{Z_F - x_n}{x_1 - x_n} < 1. \tag{139}$$

Assume that $x_n > Z_F$. Then, $x_n > x_1$, hence, $Z_F > x_1$. In that case, two easy inductions on the series of trays above and below the feed show that this is impossible:

By Equation (137),

$$j < f;\ x_{j+1} < Z_f,$$

and by Equation (138),

$$j > f,\ x_{j-1} > Z_f,$$

which give two contradictory estimations of x_f. Hence, we have the following lemma.

Lemma 1.4. *At all equilibria,* $x_1 > Z_F > x_n$.

This we already assumed in the first part of this paragraph. Therefore, $Z_F - x_n > 0,\ x_1 - x_n > 0$. Now, by (139), x_n expresses in terms of x_1, $x_n = \frac{Z_F - \beta x_1}{1-\beta}$. It follows, because we look for $x_1, x_n \in [0, 1]$, that x_1 is between Z_F and $\frac{Z_F}{\beta}$. Also, when x_1 increases from Z_F to $\frac{Z_F}{\beta}$, x_n decreases from Z_F to zero.

Using this, and (137) and (138) again, it is easy to prove by induction that we get two estimates $x_{f1}(x_1)$, $x_{f2}(x_1)$ of x_f, $x_{f1} : [Z_F, \frac{Z_F}{\beta}] \to [x-, x+]$, $x- < Z_F$, and $x_{f2} : [Z_F, \frac{Z_F}{\beta}] \to [0, y+]$, $y+ > Z_F$, $x_{f1}(x_1)$ has a strictly positive derivative, $x_{f2}(x_1)$ has a strictly negative derivative. It follows with the implicit function theorem that $(x_1, \ldots, x_n)$ express smoothly in terms of (L, V, Z_F).

Proposition 1.5. *There is a smooth mapping* $Eq : \tilde{D} \to [0, 1]^n$, $(L, V, Z_F) \to (x_1, \ldots, x_n)$, *which to* (L, V, Z_F) *associates the corresponding* ***unique*** *equilibrium* $(x_1, \ldots, x_n) \in [0, 1]^n$. *Here,* $\tilde{D} = \{(L, V, Z_F) |\ V - L > 0,\ 1 > \beta = \frac{V-L}{F}, 1 > Z_F > 0\}$.

Exercise 1.5. Show the details of the proof of Proposition 1.5.

In particular, for $(x_1, x_n) \in D$, and **if Z_F is medium** (i.e., $Z_F \in [0.2, 0.8]$), **there is a unique equilibrium (x, L, V) corresponding to (x_1, x_n), and no other equilibrium $x^* \in [0, 1]^n$ corresponds to (L, V).**

Remark 1.2. Moreover, it is easy to check that no equilibrium x^* lies in the boundary $\partial([0, 1]^n)$.

1.2.3. Asymptotic Stability

Let us fix constant values of the control, $L = L^*$, $V = V^*$. (Of course, we assume $V^* > 0$, $L^* > 0$, $V^* - L^* > 0$, $L^* + F - V^* > 0$). Let us rewrite the equation of the column, for this constant control u^* as a vector field, $\dot{x} = f(x)$. We know by the previous paragraph that we have a single equilibrium $x^* \in]0, 1[^n$, $f(x^*) = 0$.

Exercise 1.6. Show that the linearized vector field $Tf(x^*)$ at the equilibrium is asymptotically stable.

For hints, consult the paper by Rosenbrock [55].

We will now construct a locally Lipschitz, but nondifferentiable Lyapunov function $V_R(x)$ for the distillation column.

Let us set

$$V_R(x) = \sum_{i=1}^{n} |f_i(x)|. \tag{140}$$

As we shall see in our case, this nondifferentiable Lyapunov function will be sufficient for our purposes. The derivative along trajectories has to be replaced by the right-hand derivative.

Here, the right-hand derivative of V_R along the trajectories, $D^+(V_R, f) = \lim_{t\to 0^+} \frac{V_R(Exp(tf(x))) - V_R(x)}{t}$ always exists is given by

$$D^+(V_R, f) = \sum_{k=1}^{n} \alpha_k \dot{f}_k(x) = \sum_{k=1}^{n} \alpha_k L_f f_k(x), \tag{141}$$

where α_k is defined by

$$\alpha_k = \begin{pmatrix} +1, \; f_k > 0 \text{ or } (f_k = 0 \text{ and } \dot{f}_k > 0), \\ 0, \; f_k = 0 \text{ and } \dot{f}_k = 0, \\ -1, \; f_k < 0 \text{ or } (f_k = 0 \text{ and } \dot{f}_k < 0). \end{pmatrix}. \tag{142}$$

Obviously, V_R is ≥ 0, $V_R(x) = 0$ iff $= x = x^*$.

Lemma 1.6. *V_R is a Lipschitz Lyapunov function for the distillation column, i.e., $D^+(V_R, f) \leq 0$. Moreover, $D^+(V_R, f)(x) = 0$ implies $f_1(x) = f_n(x) = 0$.*

Proof. In this proof, and only in it, let us make the change of variables $H_i x_i \to x_i$.

Let us set $\psi_i = -\sum_{k=1}^{n} \frac{\partial f_k}{\partial x_i}$. A direct computation shows that we have the following properties for the distillation column, on the cube $[0, 1]^n$:

$$\psi_1 > 0, \ \psi_n > 0, \ \psi_i = 0, n > i > 1,$$
$$\frac{\partial f_i}{\partial x_i} \leq 0, i = 1, \ldots, n, \quad \frac{\partial f_k}{\partial x_i} \geq 0, k \neq i. \tag{143}$$

A trajectory $x(t)$ being fixed, let us set $G = \int_0^t \sum_{k=1}^{n} |\psi_k f_k| dt$. By Theorem 1.1, x is defined on $[0, +\infty[$, if $x_0 \in [0, 1]^n$. Then, $\frac{dG}{dt}$ exists and is ≥ 0 for all $t \geq 0$.

Let us compute $A(t) = \frac{dG}{dt} + D^+(V_R, f)$,

$$A(t) = \sum_{k=1}^{n} \alpha_k \dot{f}_k + \sum_{k=1}^{n} |\psi_k f_k|,$$

where α_k is given by (142). Therefore,

$$A(t) = \sum_{i=1}^{n} f_i \sum_{k=1}^{n} \alpha_k \frac{\partial f_k}{\partial x_i} + \sum_{k=1}^{n} |\psi_k f_k|. \tag{144}$$

Otherwise, $f_i \ \alpha_i \frac{\partial f_i}{dx_i} = -|f_i|| \frac{\partial f_i}{dx_i}|$ (by (142), and (143)). Hence, by the definition of ψ_i and by (143):

$$f_i \alpha_i \frac{\partial f_i}{\partial x_i} = -|f_i| \left(|\psi_i| + \sum_{k=1}^{n} \left\{ \left| \frac{\partial f_k}{\partial x_i} \right| \right\}_{k \neq i} \right). \tag{145}$$

Also,

$$f_i \sum_{k=1}^{n} \left\{ \alpha_k \frac{\partial f_k}{\partial x_i} \right\}_{k \neq i} \leq |f_i| \sum_{k=1}^{n} \left\{ \left| \frac{\partial f_k}{\partial x_i} \right| \right\}_{k \neq i}. \tag{146}$$

By replacing (145) and (146) in (144), we get

$$A(t) \leq \sum_{i=1}^{n} \left\{ -|f_i| \sum_{k=1}^{n} \left\{ \left| \frac{\partial f_k}{\partial x_i} \right| \right\}_{k \neq i} - |f_i \psi_i| + |f_i| \sum_{k=1}^{n} \left\{ \left| \frac{\partial f_k}{\partial x_i} \right| \right\}_{k \neq i} \right.$$
$$\left. + |f_i \psi_i| \right\} = 0.$$

Therefore,

$$D^+(V_R, f) \leq -\frac{dG}{dt} \leq 0. \tag{147}$$

By (143), $\frac{dG}{dt}$ is zero iff $f_1 = f_n = 0$. ■

We can now apply the usual reasoning stating that $x(t)$ tends to an invariant set contained in $\{D^+(V_R, f) = 0\}$. Let us check that it works also in our nonsmooth situation. Let $x(t)$ be a semitrajectory starting from $x_0 \in [0, 1]^n$. This semitrajectory being bounded, its ω-limit set $\bar{\Omega}$ is nonempty. Let $\bar{x}$ be in $\bar{\Omega}$. Because V_R is Lipschitz, V_R is almost everywhere differentiable. Hence, $D^+(V_R, f)$ is almost everywhere the derivative of $V_R(x(t))$. It is bounded and $V_R(x(t))$ is absolutely continuous. Hence,

$$\begin{aligned} V_R(x(t)) - V_R(x_0) &= \int_0^t (D^+(V_R, f)(x(\theta))d\theta \\ &\leq -\int_0^t \dot{G}(x(\theta))d\theta, \end{aligned}$$

and,

$$V_R(\bar{x}) - V_R(x_0) \leq -\int_0^\infty \dot{G}(x(\theta))d\theta.$$

Now, $\dot{G}(x) \geq 0$ is a continuous function of x. Assume that $\dot{G}(\bar{x}) > 0$.

Then we can find a closed ball $B_r(\bar{x})$, of radius r, centered at $\bar{x}$, such that $\dot{G}(z) > \delta > 0$ for all $z \in B_r(\bar{x})$.

Let $t_k \to \infty$ be a sequence such that $x(t_k) \to \bar{x}$. For k large enough, $x(t_k) \in B_{\frac{r}{2}}(\bar{x})$. Because $\|\dot{x}\|$ is bounded, $x(t)$ stays an infinite time in $B_r(\bar{x})$, and

$$\int_0^\infty \dot{G}(x(\theta))d\theta = +\infty.$$

Therefore, $\dot{G}(\bar{x}) = 0$.

The ω-limit set $\bar{\Omega}$ being invariant, if $\bar{x}(t)$ is the semitrajectory starting from $\bar{x}$, $\dot{G}(\bar{x}(t)) = 0$. This implies that $f_1(\bar{x}(t)) = f_n(\bar{x}(t)) = 0$ for all $t \geq 0$. Therefore, $\bar{x}_1(t)$ is constant, $\frac{d\bar{x}_1(t)}{dt} = 0 = V(\bar{y}_2(t) - \bar{x}_1(t))$ implies that $\bar{y}_2(t)$, and hence $\bar{x}_2(t)$ are constant. By induction, $\bar{x}_1(t), \ldots, \bar{x}_f(t)$ are constant. A similar reasoning starting from $\bar{x}_n(t) =$ constant shows that, in fact, $\bar{x}(t) =$ constant $= \bar{x}$. Therefore, $\bar{x} = x^*$ because the equilibrium is unique. Hence, x^* is a globally asymptotically stable equilibrium in restriction to $[0, 1]^n$ (we already know that it is asymptotically stable).

Theorem 1.7. *x^* is a globally asymptotically stable equilibrium for the equations of the distillation column in restriction to $[0, 1]^n$ (if $L(t) \equiv L^*$, $V(t) \equiv V^*$, $Z_F(t) \equiv Z_F^*$).*

1.2.4. Observability

Recall that the feed composition Z_F of the column is assumed to be an unknown constant. Therefore, we have to consider the state space $\Omega_e = [0, 1]^{n+1}$, and the extended observed system, of the form

$$\dot{x} = f_{Z_F}(x, u(t)), \dot{Z}_F = 0,$$
$$y = (x_1, x_n), \tag{148}$$

where $u(t) = (L(t), V(t))$ takes values in $U\phi, \lambda$.

We see that using Equation (130) and knowing $x_1(t)$, we can compute $y_2(t)$ almost everywhere; hence, we can compute $x_2(t)$ everywhere because $x_2(t)$ is absolutely continuous and $y_2(t) = k(x_2(t))$, where k is monotonous. This can be done provided that V is nonzero.

The same reasoning using Equation (131) shows that, if V is nonzero, we can compute $x_3(t), \ldots, x_f(t)$ for all $t \geq 0$.

In the same way, using (133) and (134), we can compute $x_{n-1}(t), \ldots, x_f(t)$ on the basis of $x_n(t)$, for all $t > 0$, as soon as $F + L > 0$. Now, we see that we can compute Z_F using the data of x_f, x_{f-1}, x_{f+1}, and Equation (132), provided that F is positive. Therefore, if the set of values of control is $U\phi, \lambda$, $\phi > 0$, $\lambda > 0$, our system is observable.

Theorem 1.8. *If $\phi > 0, \lambda > 0$, the extended system (148) is observable for all L^∞ inputs with values in $U_{\phi,\lambda}$. (The state space is $[0, 1]^{n+1}$ or $]0, 1[^{n+1}$.)*

On the contrary, if L, V, F are all zero, the system is obviously unobservable.

Exercise 1.7. Study intermediate situations, where at least one but not all variables L, V, F are identically zero.

Remark 1.3. The bad input takes values in the boundary of the physical control set $U_p = \{L \geq 0,\ F \geq 0, V - L \geq 0, F + L - V \geq 0\}$. For this input, there are indistinguishable trajectories entirely contained in the boundary of the physical state space $X_p = \{0 \leq x_i, Z_F \leq 1\}$. This seems to be a very common situation in practice.

Exercise 1.8. Prove that, in fact, if $\phi > 0$, $\lambda > 0$, the system is also uniformly infinitesimally observable.

Remark 1.4. The distillation column has the same number of outputs and inputs (2). It is observable, and even uniformly infinitesimally observable. Hence, it is nongeneric from the point of view of observability, by our theory.

1.3. Observers

To simplify the exposition, let us assume that $n = 2f - 1$. Let us consider the following embedding $\mathcal{E} : R^{2f} \to R^{2f+1}$:

$$(x_1, \ldots, x_{2f-1}, Z_F) \to X = (X^1, \ldots, X^f, Z_F),$$

with

$$\begin{aligned} X^1 &= \left(X_1^1, X_2^1\right) = (x_1, x_{2f-1}), X^i = \left(X_1^i, X_2^i\right) = (x_i, x_{2f-i}), \\ X^{f-1} &= \left(X_1^{f-1}, X_2^{f-1}\right) = (x_{f-1}, x_{f+1}), \\ X^f &= (x_f, x_f), X^{f+1} = Z_F. \end{aligned} \tag{149}$$

Let us set $\underline{X^i} = (X^1, \ldots, X^i)$. Then, the equation of the column can be rewritten as

$$\begin{aligned} \dot{X}^1 &= F^1(\underline{X^2}, L, V), \\ &\vdots \\ \dot{X}^i &= F^i(\underline{X^{i+1}}, L, V), \\ &\vdots \\ \dot{X}^{f-1} &= F^{f-1}(\underline{X^f}, L, V), \\ \dot{X}^f &= F^f(X, L, V), \\ \dot{X}^{f+1} &= 0, \\ y &= X^1. \end{aligned} \tag{150}$$

Moreover, in this equation, we have

$$\begin{aligned} \frac{\partial F^i}{\partial X^{i+1}} &= \begin{pmatrix} \alpha_i(\underline{X^{i+1}}, u), 0 \\ 0, \beta_i(\underline{X^{i+1}}, u) \end{pmatrix}, 1 \le i \le f-1, \\ \frac{\partial F^f}{\partial X^{f+1}} &= \begin{pmatrix} \alpha_f \\ \beta_f \end{pmatrix}, \end{aligned} \tag{151}$$

where $\alpha_i, \beta_i, 1 \le i \le f-1$ are strictly positive functions, and α_f, β_f are strictly positive constants. (In fact, $\beta_i(\underline{X^{i+1}}, u)$ depends only on u.) Also, **we can extend the function $k(x)$ smoothly to all of R, outside the interval $[0, 1]$, so that, for all $x \in R$:**

$$0 < \mu < k'(x) < \nu < +\infty. \tag{152}$$

Therefore, in fact, we get that Equations (150) of the column are defined on all of R^n, smooth, and

$\overline{A_2}$. If u takes values in $U_{\phi,\lambda}$, $\phi, \lambda > 0$, there are positive constants A and B, such that $0 < A \leq \alpha_i, \beta_i \leq B$,

$\overline{A_1}$. the F^i in (150) are globally Lipschitz on R^{n+2}, with respect to $\underline{X^i}$, uniformly with respect to u and X^{i+1}.

Let us denote by A, C the following block matrices:

$$A = \begin{pmatrix} 0, \dfrac{\partial F^1}{\partial X^2}, 0, \ldots\ldots\ldots\ldots, 0 \\ \cdot \\ \cdot \\ 0, 0, \ldots\ldots\ldots, 0, \dfrac{\partial F^{f-1}}{\partial X^f}, 0 \\ 0, \ldots\ldots\ldots\ldots, 0, \dfrac{\partial F^f}{\partial X^{f+1}} \\ 0, \ldots\ldots\ldots\ldots\ldots\ldots, 0 \end{pmatrix}, \qquad (153)$$

$$C = (Id_2, 0, \ldots\ldots\ldots\ldots\ldots, 0).$$

$A : R^{2f+1} \to R^{2f+1}$, $C : R^{2f+1} \to R^2$.

Along a trajectory $X(t)$ of the system, set $A = A(t)$.

Exercise 1.9. Prove that Lemma 2.1 of Chapter 6 generalizes to this case, i.e., there exist $S, K, \tilde{\lambda}, \tilde{\lambda} > 0$, $K : R^2 \to R^{n+2}$, S, symmetric positive definite, such that $(A(t) - KC)'S + S(A(t) - KC) \leq -\tilde{\lambda}\, Id$. Moreover, $S, K, \tilde{\lambda}$ depend only on ϕ, λ (the parameters of the set of values of control $U_{\phi,\lambda}$).

In this exercise, the special form of the 2×2 matrices $\frac{\partial Fi}{\partial X^{i+1}}$ is crucial.

Now, Theorem 2.2 and Corollary 2.3 of Chapter 6 also generalize. Consider the system

$$(\Sigma_O)\ \frac{d\hat{X}}{dt} = F(\hat{X}, u) - K_\theta(C\hat{X} - y(t)), \qquad (154)$$

with $K_\theta = \Delta_\theta K$,

$$\Delta_\theta = \begin{pmatrix} \theta Id_2, 0, \ldots\ldots, 0 \\ \cdot \\ \cdot \\ 0, \ldots, 0, \theta^f Id_2, 0 \\ 0, \ldots\ldots, 0, \theta^{f+1} \end{pmatrix}.$$

Theorem 1.9. *The system (154) is an exponential state observer for all L^∞ inputs with values in $U_{\phi,\lambda}$, relative to $\{(x, Z_F) \in]0, 1[^{n+1}\}$.*

Remark 1.5. In fact, we have more than that: (154) is an $U^{0,B}$ exponential state observer for (150) relative to any open relatively compact $\Omega \subset R^{n+2}$. However, estimations out of $[0, 1]^{n+2}$ have no physical meaning.

Exercise 1.10. Prove Theorem 1.9. (use the proof of Theorem 2.2 of Chapter 6).

There is also the possibility to generalize the high gain EKF to this case. Let us show how.

Considering the modified equation of the distillation column (150), remember that $X_1^i = x_i$, $X_2^i = x_{n-i+1}$, $i = 1, \ldots, f$. Set

$$\begin{aligned}
\tilde{X}_2^i &= X_2^i, i = 1, \ldots, f, \\
\tilde{X}_1^1 &= X_1^1, \tilde{X}_1^2 = k(X_1^2), \tilde{X}_1^3 = k'(X_1^2)k(X_1^3), \ldots, \\
\tilde{X}_1^f &= k'(X_1^2)\ldots k'(X_1^{f-1})k(X_1^f), \qquad (155)\\
\tilde{X}_1^{f+1} &= k'(X_1^2)\ldots k'(X_1^{f-1})k'(X_1^f)Z_F, \tilde{X}_2^{f+1} = Z_F, \\
\tilde{X}^i &= (\tilde{X}_1^i, \tilde{X}_2^i), \tilde{X} = (\tilde{X}^1, \ldots, \tilde{X}^{f+1}), \tilde{X} \in R^{n+3} = R^{2f+2}.
\end{aligned}$$

It is easy to check that the equation of the distillation column is now of the following form.

Lemma 1.10. *Equations (150) become*

$$\frac{d\tilde{X}}{dt} = A(u)\tilde{X} + \tilde{F}(\tilde{X}, u), \qquad (156)$$

$$y = \tilde{X}^1, \qquad (157)$$

where

$$A(u) = \begin{pmatrix} 0, U_1, 0, \ldots\ldots\ldots, 0 \\ \vdots \\ 0, \ldots\ldots\ldots, 0, U_1, 0 \\ 0, \ldots\ldots\ldots\ldots, 0, U_2 \\ 0, \ldots\ldots\ldots\ldots\ldots, 0 \end{pmatrix}, U_1 = \begin{pmatrix} V, & 0 \\ 0, & F+L \end{pmatrix},$$

$$U_2 = \begin{pmatrix} F, 0 \\ 0, F \end{pmatrix}$$

and $\tilde{F}$ is lower block-triangular: the components $\tilde{F}^{2i-1}(\tilde{X}, u)$ and $\tilde{F}^{2i}(\tilde{X}, u)$ depend only on $(\tilde{X}^1, \ldots, \tilde{X}^i)$, $1 \leq i \leq f+1$.

Moreover, $\tilde{F}^{2i-1}(\tilde{X}, u)$ and $\tilde{F}^{2i}(\tilde{X}, u)$, which are well defined on $[0, 1]^{2i}$ can be extended to all of R^{2i} as C^∞ compactly supported functions.

Exercise 1.11. Prove Lemma 1.10.

Now, the following theorem holds:

Theorem 1.11. *The high-gain extended Kalman filter construction extends (in both its continuous-continuous and continuous-discrete forms) to the systems of the form (156), provided that $u \in U_{\phi,\lambda}$, $\phi, \lambda > 0$.*

Exercise 1.12. Prove Theorem 1.11.

Unfortunately, Theorem 1.11 is only a theoretical result. Because of the complication and the poor "conditioning" of the change of variables $X \to \tilde{X}$, this last construction did not show very good performances in practice. For a reasonable version of the high gain EKF from the practical point of view, see [64].

1.4. Feedback Stabilization

Before stabilizing by using the outputs only and an observer, we have to stabilize first by state feedback. We already know from Section 1.2.3 that, given target output values $y^* = (x_1^*, x_n^*)$, there is a unique value of the controls $u^* = (L^*, V^*)$ and a corresponding unique stationary state x^*, such the distillation column is asymptotically stable within $[0, 1]^n$ at x^* if one applies as controls the constant stationary values u^*. Moreover, u^* is a smooth function of (x_1^*, x_n^*, Z_F), defined on a certain open domain $\bar{D} \subset [0, 1]^3$. Therefore, given (x_1^*, x_n^*), we have a first trivial feedback for the distillation column

$$u_1^*(Z_F) = u^*(x_1^*, x_n^*, Z_F). \tag{158}$$

This feedback will be sufficient to solve the second problem posed in Section 1.1.3. However, it does not increase the **response time** of the column.

Let us now briefly discuss how we can construct another feedback $u_2^*(Z_F, x_1, \ldots, x_n)$ for this purpose. This feedback is constructed on the basis of u_1^* using inverse Lyapunov's theorems, and the classical idea to **enhance stability**.

Let us reparametrize the controls and set $u = (u_1, u_2)$, $u_1 = L$, $u_2 = V - L$. By our assumptions on the domain $U_{\phi,\lambda}$, we can assume that $0 < u_i^{\min} \leq u_i \leq u_i^{\max}$, $i = 1, 2$, and that for $Z_F \in [.2, .8]$, $u_i^{\min} < (u_1^*)_i(Z_F) < u_i^{\max}$. For $i = 1, 2$, $(u_1^*)_i$ can be extended to a smooth function: $R \to]u_i^{\min}, u_i^{\max}[$, still denoted by $(u_1^*)_i$.

Let $\varphi_i : R \to R$ be as follows: φ_i is C^∞, increasing, is equal to $u_i^{\min}$ in a neighborhood of $-\infty$, to $u_i^{\max}$ in a neighborhood of $+\infty$, and to the identity on an interval containing the compact $(u_1^*)_i([.2, .8])$.

Let us observe also that the equations of the distillation column are **control-affine**:

$$\dot{x} = f(x) + u_1 g_1(x) + u_2 g_2(x) = f(x) + g(x)u, \tag{159}$$

and because the feed composition is unknown,

$$\dot{Z}_F = 0 \tag{160}$$

is an additional equation to be taken into account.

The feedback $u_1^*(Z_F) = (L^*, V^* - L^*)$ being a smooth function of the unknown Z_F, we can find $V_{Z_F}(x)$, a family of smooth proper Lyapunov functions defined on open neighborhoods of $[0, 1]^n$, for the vector fields $\dot{x} = f(x) + g(x)u_1^*(Z_F) = F_{Z_F}^*(x)$.

Exercise 1.13. Show that such a family $V_{Z_F}(x)$ does exist.

This exercise is not very difficult. However, it is a bit technical. Using the result of Exercise 1.6, we can construct explicitly a smooth family of local Lyapunov functions, that is, solving the linear Lyapunov equation, depending smoothly on Z_F, at each equilibrium point (which also depends smoothly on Z_F). Second, using the global asymptotic stability of each equilibrium within $[0, 1]^n$, one can modify the arguments in the classical proof of the global Lyapunov's inverse theorems.

For $R \geq 0$, let us define the feedback $u_2^*(Z_F, x)$ as follows:

$$(u_2^*)_i(Z_F, x) = \varphi_i(-RL_{g_i} V_{Z_F}(x) + (u_1^*)_i(Z_F)), \ i = 1, 2. \tag{161}$$

If $R = 0$, the smooth feedback u_2^* is just the feedback u_1^*. If $R > 0$, then the derivative of V_{Z_F} along the trajectories of the column is more negative than the derivative along the trajectories corresponding to the (constant) feedback $u_1^*(Z_F)$, by construction. Hence, with this feedback u_2^*, the column is also globally asymptotically stable, and in practice we get a faster convergence.

It is difficult to handle the Lyapunov function V_{Z_F} in practice. For a construction of another feedback based on the same idea but using the (singular)

Lyapunov function V_R in place of V_{Z_F}, see Ref. [64]. (One cannot use directly this Lyapunov function, because it leads to discontinuous feedbacks. But, it is possible to smooth these feedbacks.)

1.5. Output Stabilization

We can use the feedback $u_1^*(Z_F)$ coupled to the high-gain observer (Σ_O) of Formula (154). Again, we set $\hat{X} = (\hat{x}, \hat{Z}_F) \in R^{n+2}$ for the vector of estimates of $x = (x_1, \ldots, x_n)$ and Z_F. Then, $\hat{X} = (\hat{x}, \hat{Z}_F) = \mathcal{E}(\hat{x}_1, \ldots \hat{x}_n, \hat{Z}_F)$, where $\mathcal{E}$ is the mapping defined in Section 1.3.

We get a set of equations of the form

$$\begin{aligned} &(i)\quad \frac{d\hat{X}}{dt} = F(\hat{X}, u_1^*(\hat{Z}_F)) - K_\theta(C\hat{X} - y(t)), \\ &(ii)\quad \frac{dx}{dt} = \tilde{F}(x, u_1^*(\hat{Z}_F)). \end{aligned} \tag{162}$$

Under this control $u_1^*(\hat{Z}_F)(t)$, we know that x stays in $[0, 1]^n$ because $(u_1^*)_i(\hat{Z}_F)(t) \in]u_i^{\min}, u_i^{\max}[$(see Section 1.2.1).

Any semitrajectory $\{x(t); t \geq 0\}$ is bounded (stays in $[0, 1]^n$). Set $\varepsilon = \hat{X} - \mathcal{E}(x, Z_F)$. By Theorem 1.9, ε goes exponentially to zero. Hence, any semitrajectory $\{\zeta(t) = (\hat{X}(t), x(t)); t \geq 0\}$ of (162) starting from $R^{n+2}\times [0, 1]^n$, is bounded in R^{2n+2}. The ω-limit set Ω of $\zeta(t)$ is nonempty. The set $\{\varepsilon = 0\}$ is an invariant manifold, and the dynamics on $\{\varepsilon = 0\}$ is exponentially stable (Exercise 1.6). Therefore, the whole system (2) is asymptotically stable within $R^{n+2}\times [0, 1]^n$, with the usual arguments.

Theorem 1.12. *The distillation column, with the feedback $u_1^*(Z_F)$ taken on the estimate $\hat{Z}_F$ of Z_F given by the high-gain observer of Theorem 1.9, is globally asymptotically stable (asymptotically stable within $R^{n+2} \times [0, 1]^n$).*

Exercise 1.14. Show a similar theorem with the high-gain extended Kalman filter of Theorem 1.11. (Consider both the continuous-continuous and continuous-discrete forms.)

Exercise 1.15. Show the same results using the feedback u_2^*.

Remark 1.6. Observe that any (small) exponential rate of convergence of the observer is sufficient. We are in the propitious situation of Section 3 in the previous chapter (Chapter 7).

2. Polymerization Reactors

This second application concerns "polymerization reactors." The main difference with the previous example of distillation columns is that the "output stabilization problem" deals with (exponentially) unstable equilibria. In practice, we are concerned with **styrene** polymerization, which is very common. For details, we refer to paper [65] and to thesis [63].

2.1. The Equations of the Polymerization Reactor

The kinetics of radical polymerization is well known. Here, we consider a "stirred tank reactor" in which "free radical polymerization" takes place. The reactor is continuously fed by monomer, initiator, and solvent. The growth of monomer molecules into polymer chains is induced by "free radicals," which are generated by the initiator decomposition.

Schematically, a certain number of reactions occur, as: initiator decomposition, initiation, propagation, chain transfer, and chain termination.

A certain number of more or less realistic assumptions are made, that we mention for people who know something about polymerization: equal reactivity, quasi steady-state for all radical species.

The equations are obtained just by writing a mass balance of monomer, solvent, and initiator. The polymer is obtained by difference.

The variables W_M, W_S, W_I denote the weight fractions of monomer, solvent, and initiator in the reactor. The variables W_{MF}, W_{SF}, W_{IF} denote the same weight fractions in the feed of the reactor. We get the usual equations.

Mass balance:

$$\begin{aligned}
&(1)\quad \frac{dW_M}{dt} = \frac{Q_{MF}}{\rho V}(W_{MF} - W_M) - (k_p + k_{ttm})W_M C_R - 2f\, k_d W_I \frac{M_M}{M_I},\\
&(2)\quad \frac{dW_S}{dt} = \frac{Q_{MF}}{\rho V}(W_{SF} - W_S) - k_{tts} W_S C_R, \qquad\qquad (163)\\
&(3)\quad \frac{dW_I}{dt} = \frac{Q_{MF}}{\rho V}(W_{IF} - W_I) - k_d W_I.
\end{aligned}$$

Here Q_{MF} is the mass flow of the feed, V is the volume of the reactor, C_R is the total concentration of radical chains, and f is the "initiator efficiency" (a constant). The constants M_M, M_S, M_I are the molecular weights of monomer, solvent, and initiator. The variables $k_p, k_{ttm}, k_{tts}, k_d$ are the kinetic rates, and they depend on the temperature T in the reactor. (The subscripts $p, ttm, \ldots$ mean "propagation", "transfer to monomer", ...).

Moreover, we have

$$\text{(a)}\ \frac{1}{\rho}=\left(\frac{W_M}{\rho_M}+\frac{W_S}{\rho_S}+\frac{W_P}{\rho_P}\right)=\left(\frac{W_M}{\rho_M}+\frac{W_S}{\rho_S}+\frac{1-W_M-W_S}{\rho_P}\right),$$
$$\text{(b)}\ C_R=\sqrt{\frac{f\,k_d C_I}{k_{tc}+k_{td}}},\quad C_I=\frac{\rho W_I}{M_I}. \tag{164}$$

Here, ρ_M, ρ_S, ρ_P are the densities of monomer, solvent, and polymer (constants). The variable C_I is the molar concentration of initiator. Additionally, we have the thermal balance, expressing the heat exchange with a cooling flow.

Heat balance:

$$\frac{dT}{dt}=\frac{Q_{MF}}{\rho V}\left(\frac{C_{PF}}{C_P}T_F-T\right)+\frac{U}{\rho V C_P}(T_{heat}-T)-\frac{k_p C_R W_M \Delta H}{M_M C_P}. \tag{165}$$

The variables C_P, C_{PF} are the heat capacities of the reacting mixture and the feed (assumed to be constants, for simplicity). The feed temperature is denoted by T_F, and U is the heat transfer coefficient (both are constants), T_{heat} is the temperature of the cooling flow (one of the control variables, later on). Also, Δ_H denotes the global reaction heat for the polymerization reaction (a negative constant). The next (and last) step is the **characterization of polymer chains.**

The length of polymer chains (the quality of the products of the reaction; the length of the polystyrene molecules in our case) is usually characterized by the leading moments

$$\lambda_0=\frac{1}{\rho}\sum_{x=1}^{\infty}C_{Px},\ \lambda_1=\frac{1}{\rho}\sum_{x=1}^{\infty}xC_{Px},\ \lambda_2=\frac{1}{\rho}\sum_{x=1}^{\infty}x^2C_{Px}. \tag{166}$$

Here, x is the length chain and C_{Px} is the concentration in the reacting mixture of **dead polymer** of length x. One has, by difference on the mass balance,

$$\lambda_1=\frac{1-W_M-W_S}{M_M}. \tag{167}$$

In place of λ_0, λ_2, chemical engineers prefer to consider the following two quantities, M_n and PD, called respectively the "number average molecular weight" and the "polydispersity":

$$M_n = M_M \frac{\lambda_1}{\lambda_0} = \frac{1 - W_M - W_S}{\lambda_0},$$
$$PD = \frac{\lambda_0 \lambda_2}{(1 - W_M - W_S)^2} M_M^2. \tag{168}$$

These three quantities, λ_1, M_n, and PD, characterize the quality of the polymer produced by the reactor. As we saw, λ_1 is a direct expression of the state variables W_M, W_S of the set of Equations (163). The variables λ_0 and λ_2 satisfy

$$\frac{d\lambda_0}{dt} = \frac{Q_{MF}}{\rho V}(\lambda_{0F} - \lambda_0) + \frac{\Phi_0}{\rho},$$
$$\frac{d\lambda_2}{dt} = \frac{Q_{MF}}{\rho V}(\lambda_{2F} - \lambda_2) + \frac{\Phi_2}{\rho}, \tag{169}$$

where λ_{0F} and λ_{2F} are constants, and Φ_0 and Φ_2 are smooth functions of the state variables (W_M, W_S, W_I, T) in (163) and (165). For the detailed expressions of Φ_0, Φ_2, one can consult [63] or [65].

2.2. Assumptions and Simplifications of the Equations

1. All the kinetic constants ($k_d, k_{ttm}, \ldots$), depending on the temperature of the reaction, have the Arrhenius form

$$k = k_0 \, e^{-\frac{E}{RT}}. \tag{170}$$

2. Engineers usually want to go to a certain equilibrium by acting on the two control variables: $u_1 = W_{IF}$ and $u_2 = T_{heat}$. Then, u_1 is the initiator concentration in the feed, and u_2 is the cooling temperature. All the other variables are constant (such as W_{MF} and W_{SF}, fractions of monomer and solvent in the feed). They depend on the target steady-state only.

Because the control is two-dimensional, $u = (u_1, u_2)$, it is reasonable to fix two output variables at the steady states. Usually, they are $\bar{y}_1 = T$, the reaction temperature, and $\bar{y}_2 = \chi = 1 - \frac{W_M}{W_{MF}}$. In practice, χ is called the "conversion rate." It represents the ratio of monomer actually converted in polymer.

The point is that, here the observations will not be the same variables as the outputs to be controlled: They are: $y = (y_1, y_2) = (T, \frac{1}{\rho})$. So that, we are in the situation of a two inputs, two outputs (to be driven to a "set value"), two observations, control system.

The state is 6-dimensional:

$$\bar{x} = (W_M, W_S, W_I, T, \lambda_0, \lambda_2).$$

The six equations are the mass balance (163), the heat balance (165), and the equations of the moments (169). In fact, this set of six equations is **triangular**, as one can see. The evolution of the first part of the state $x = (W_M, W_S, W_I, T)$ does not depend on λ_0, λ_2. Otherwise, the two output variables, $\bar{y}_1 = T$ and $\bar{y}_2 = \chi$, and also the observed variables, $y_1 = T$ and $y_2 = \frac{1}{\rho}$ do not depend on λ_0, λ_2. This means that λ_0, λ_2 are "unobservable" variables. We can just **forget them**: This means that we quotient the system by the "trivial foliation" Δ_Σ, which is not trivial in that case (see Section 5 of Chapter 2). If we don't care about the behavior of λ_0, λ_2, this is perfectly correct.

In fact, it happens that after asymptotically stabilizing the first part x of the state, the second part $\lambda = (\lambda_0, \lambda_2)$ goes to a uniquely determined equilibrium. This follows from the set of equations describing the reactor: If we have a stabilizing feedback for the first part x of the state, then the full system is also asymptotically stable (this is not obvious, but can be proven).

Exercise 2.1. Prove the last statement (making reasonable assumptions on the functions Φ_0, Φ_2).

Therefore, **the equations of the reactor reduce to the set of four equations for** x, **given by** (163) **and** (165).

The full treatment of these equations with our theory is shown in [65]. However, it is a bit tricky. Here, for the sake of clarity and simplicity of exposition, **we will make one more simplification**, reasonable in the case of the styrene reactor.

We will assume that Equation (163), part (1) reduces to

$$(1')\ \frac{dW_M}{dt} = \frac{Q_{MF}}{\rho V}(W_{MF} - W_M) - k_p W_M C_R. \tag{171}$$

It means that in Equation (163, 1), the effect of the kinetic coefficients k_{ttm} and k_d (transfer to monomer and initiator decomposition) is negligible with respect to that of k_p (propagation). It is what is expected of the reaction: to produce long chains. See [63] for details. Therefore, at the end, we get the final set of equations that we will consider to describe the polymerization reactor:

$$
\begin{aligned}
&(1)\ \frac{dW_M}{dt} = \frac{Q_{MF}}{\rho V}(W_{MF} - W_M) - k_p W_M C_R, \\
&(2)\ \frac{dW_S}{dt} = \frac{Q_{MF}}{\rho V}(W_{SF} - W_S) - k_{tts} W_S C_R, \\
&(3)\ \frac{dW_I}{dt} = \frac{Q_{MF}}{\rho V}(W_{IF} - W_I) - k_d W_I, \\
&(4)\ \frac{dT}{dt} = \frac{Q_{MF}}{\rho V}\left(\frac{C_{PF}}{C_P}T_F - T\right) + \frac{U}{\rho V C_P}(T_{heat} - T) \\
&\qquad - \frac{k_p C_R W_M \Delta H}{M_M C_P}.
\end{aligned}
\tag{172}
$$

The additional relations (164) give the expressions of ρ and C_R in terms of the other variables (W_M, W_S, W_I, T). The control variables are $u = (u_1, u_2) = (W_{IF}, T_{heat})$, the variables to be controlled (i.e., driven to an expected stationary value) are $\bar{y} = (\bar{y}_1, \bar{y}_2) = (W_M, T)$, and the observed variables are $y = (y_1, y_2) = (T, \frac{1}{\rho})$. The state $x = (W_M, W_S, W_I, T)$ belongs to $X = [0, 1]^3 \times R^+$.

The equations of the reactor are control affine:

$$\dot{x} = f(x) + g_1(x)u_1 + g_2(x)u_2. \tag{173}$$

In practice, the assumption of a continuous measurement for T and ρ is reasonable.

2.3. The Problems to Be Solved

First Problem. Observation (estimation of the full state x).

Second problem. Given target values for $\bar{y}$, asymptotically stabilize the reactor at an equilibrium x^* corresponding to $\bar{y}$. Do this using the observations y only. If possible, fix the "response time."

Depending on the operating conditions (mainly depending on χ, the "monomer conversion rate"), the reactor has equilibria with different behaviors:

1. Low monomer conversion rate χ. These equilibria are (locally) asymptotically stable (exponentially).

2. High monomer conversion rate χ. These equilibria also are locally asymptotically stable. From the economic point of view, they would be the most interesting because χ just expresses the quantity of monomer that is transformed into polymer. However, they are not reasonable targets, because catastrophic phenomena occur, such as solidification, due to the high viscosity of the reacting mixture.
3. Medium monomer conversion rate χ. Unfortunately, these equilibria are unstable.

Our theory shows the possibility of stabilizing asymptotically within large domains at these equilibria of type (3). This could be very interesting in practice. In practice, the users are already much interested by solving Problem 1 (just observation).

2.4. Properties of the Equations (172)

2.4.1. The Equilibria

The target values of W_M and $T > 0$ being chosen, at equilibria, Equations (172) give

$$(1)\ 0 = W_{MF} - W_M - \frac{k_p W_M V}{Q_{MF}} \rho C_R.$$

Here k_p is a function of T, hence, ρC_R is determined uniquely by the knowledge of T and W_M. If $0 < W_M < W_{MF}$ (which will be the case), then $\rho C_R > 0$.

$$(4)\ 0 = \frac{C_{PF}}{C_P} T_F - T + \frac{U}{C_P Q_{MF}} (T_{heat} - T) - \frac{k_p W_M \Delta_H V}{M_M C_P Q_{MF}} \rho C_R.$$

Now, ρC_R, W_M, T being known, T_{heat} is determined uniquely (k_p is a function of T, and other variables are constants).

$$(2)\ 0 = W_{SF} - W_S - \frac{k_{tts} V}{Q_{MF}} \rho C_R W_S.$$

This equation determines W_S uniquely, because $\rho C_R > 0$. We find $0 < W_S < W_{SF}$.

Now, W_S and W_M being known, ρ is determined uniquely by Equation (164a), and hence C_R is determined uniquely. Then, $W_I > 0$ is determined uniquely by (164b).

$$(3)\ 0 = W_{IF} - W_I - \frac{k_d \rho V}{Q_{MF}} W_I.$$

This last equation determines uniquely the control W_{IF}. We have shown the following lemma.

Lemma 2.1. *If the target stationary value $\bar{y}^*$ of $\bar{y} = (W_M, T)$ is given, then there is a unique corresponding equilibrium x^* of x and a unique corresponding value u^* of the stationary control $u = (W_{IF}, T_{heat})$.*

Moreover, for reasonable values of all the other constants and of $\bar{y}^*$, the corresponding values of x^* are reasonable (i.e., W_M, W_S, W_I, W_{IF} are between 0 and 1, T and T_{heat} are larger than 0, ρ is $> 0, \ldots$).

For instance, let us chose a target stationary value, **corresponding to a medium conversion rate**, i.e., the corresponding equilibrium is unstable, as we shall see. To do this, we have to give the values of all constants in the equations:

$$\begin{aligned}
R &= 8.32, \; k_{td} = 0, \\
k_d &= 1.58\ 10^{15} e^{-128.10^3/(RT)}, \; k_p = 1.051\ 10^4 e^{-29.54\ 10^3/(RT)}, \\
k_{tts} &= 5.58\ 10^5 e^{-71.8\ 10^3/(RT)}, \; k_{tc} = 0.6275\ 10^6 e^{-7.026\ 10^3/(RT)}, \\
Q_{MF} &= 0.1, \; V = 1, \; f = 0.6, \; M_I = 0.164, \; M_M = 0.104, \; \rho_M = 830, \\
\rho_S &= 790, \; \rho_P = 1025, \; U = 950, \; W_{MF} = 0.9, \; W_{SF} = 0.1, \\
C_P &= 1855, \; C_{PF} = 1978, \; T_F = 288.15, \; \Delta H = -74400. \qquad (174)
\end{aligned}$$

Let us chose the target values $\bar{y}^* = (W_M^*, T^*)$ as

$$\bar{y}^* = (W_M^*, T^*) = (0.7, 330). \qquad (175)$$

Then, the computations of Lemma 2.1 above give the corresponding values for the stationary state x^* and the stationary control $u^* = (W_{IF}^*, T_{heat}^*)$:

$$\begin{aligned}
x^* &= (W_M^*, W_S^*, W_I^*, T^*) = (0.7, 0.0999997, 0.03837, 330), \\
u^* &= (W_{IF}^*, T_{heat}^*) = (0.0413, 319.38). \qquad (176)
\end{aligned}$$

Now, linearizing Equations (172) of the reactor at this equilibrium leads to the following set of eigenvalues:

$$\begin{aligned}
E_1 = \{&-0.0001238, 0.00001757 + 0.00005499\sqrt{-1}, \\
&0.00001757 - 0.00005499\sqrt{-1}, -0.0001165\}, \qquad (177)
\end{aligned}$$

which shows that, as we said, this is an unstable equilibrium. Here, the time unit is the second, which explains these low values.

The following observations will be crucial: Let us consider the equations of the reactor (172) at **isothermal and isoinitiator operation,** i.e., the two first equations in which the temperature is constrained at the value T^*, the equilibrium temperature, and W_I is constrained at the value W_I^*, the equilibrium initiator concentration.

Linearizing this set of two equations at the equilibrium gives the following second set of eigenvalues:

$$E_2 = \{-0.0001429, -0.0001165\}.$$

Thus, the following holds:

Observation 1: The target outputs W_M^*, T^* are fixed to the above values. In isothermal and isoinitiator operation, the reactor is asymptotically stable at the corresponding equilibrium.

Also, we can check on the linearized equations that the quadratic function

$$V_L(W_M, W_S) = \frac{1}{2}((W_M - W_M^*)^2 + (W_S - W_S^*)^2) \tag{178}$$

is a quadratic Lyapunov function for the linearized Equations (172, (1) and (2)), (in isothermal and isoinitiator operation).

In fact, there is more than that, as we shall see immediately. First, let us go back to the complete equations of the reactor in isothermal and isoinitiator operation:

$$\begin{aligned} (1)\quad \frac{dW_M}{dt} &= \frac{Q_{MF}}{\rho V}(W_{MF} - W_M) - k_p W_M C_R, \\ (2)\quad \frac{dW_S}{dt} &= \frac{Q_{MF}}{\rho V}(W_{SF} - W_S) - k_{tts} W_S C_R, \end{aligned} \tag{179}$$

with

$$\begin{aligned} &(a)\quad \frac{1}{\rho} = \left(\frac{W_M}{\rho_M} + \frac{W_S}{\rho_S} + \frac{W_P}{\rho_P}\right) = \left(\frac{W_M}{\rho_M} + \frac{W_S}{\rho_S} + \frac{1 - W_M - W_S}{\rho_P}\right), \\ &(b)\quad C_R = \sqrt{\frac{f\, k_d C_I}{k_{tc} + k_{td}}},\quad C_I = \frac{\rho W_I}{M_I}, \end{aligned} \tag{180}$$

and $T = T^*$, $W_I = W_I^*$.

It is easy to check that the domain,

$$\Omega_{ISO} = \{(W_M, W_S); 0 \leq W_M \leq W_{MF}, 0 \leq W_S \leq W_{SF}\}, \tag{181}$$

is a positively invariant domain for (179). Also, one can consider the quadratic function $V_L(W_M, W_S)$ defined in (178) above.

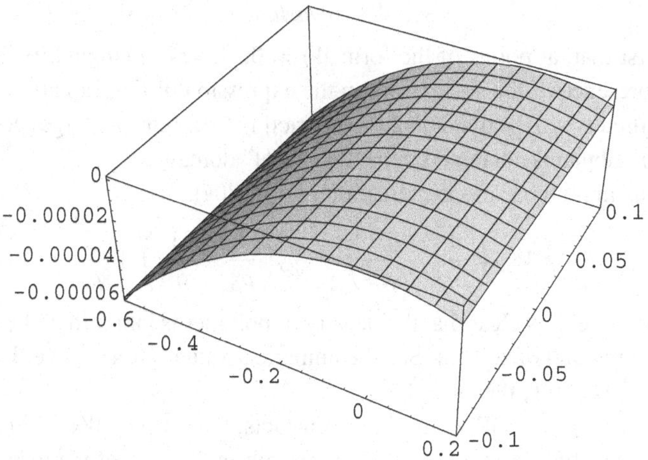

Figure 2. The derivative of V_L along the trajectories, in isothermal and isoinitiator operation.

Observation 2: The target outputs W_M^*, T^* are fixed to the above values. In isothermal and isoinitiator operation, the function V_L, which is a Lyapunov function of the linearized at the equilibrium, **is also a Lyapunov function of the nonlinearized system, in restriction to the domain Ω_{ISO}**. (Here, by "Lyapunov function, we mean strict Lyapunov function", i.e., the derivative of V_L along trajectories is strictly negative, except for the equilibrium point trajectory.)

Unfortunately, to show that this observation holds requires numerical investigation (easy). It is the reason why we gave the values (174) of all constants in the equations of the reactor. Figure 2 shows the graph of the function $\frac{dV_L}{dt}(x, y)$, where $x = W_M - W_M^*$, $y = W_S - W_S^*$, in restriction to the domain Ω_{ISO}. Therefore, using the fact that Ω_{ISO} is a compact positively invariant domain, we can state the following proposition:

Proposition 2.2. *The target outputs W_M^*, T^* are fixed to the above numerical values.* ***The equations of the reactor in isothermal and isoinitiator operation are asymptotically stable within Ω_{ISO}.***

Remark 2.1. Notice that to prove this proposition, one has to use Lasalle's invariance principle: The function V_L is not proper on the open domain $Int(\Omega_{ISO})$.

2.4.2. Invariance

Note first that, at points of the form $W_I = 0$, there is a singularity, because of the presence of the square root in the expression of C_R, Equation (180b). This difficulty will be overcomed by restricting to (arbitrarily large) positively invariant domains contained in the "physical" domain.

Let us first go back to the equation of ρ, (180a):

$$\frac{1}{\rho} = W_M\left(\frac{1}{\rho_M} - \frac{1}{\rho_P}\right) + W_S\left(\frac{1}{\rho_S} - \frac{1}{\rho_P}\right) + \frac{1}{\rho_P}. \tag{182}$$

In practice, it is clear that the density of polymer is larger than the density of monomer and of solvent. See the numerical values (174) above. Hence, if $0 \leq W_M, W_S \leq 1$, then $\rho > 0$.

Now, W_{MF}, and W_{SF} are fixed constants, $0 < W_{MF}, W_{SF} < 1$. Let us consider for the restricted state space any subset $\Omega \subset R^4$ of the form

$$\Omega = [0, W_{MF}] \times [0, W_{SF}] \times [\varepsilon_I, 1] \times [T_{\min}, T_{\max}], \tag{183}$$

and for the restricted set of values of the control $u = (W_{IF}, T_{heat})$, a set U of the form

$$U = [\varepsilon_1, 1] \times [T_{\min}, T_{\max}],$$

with $(P) : \varepsilon_I, \varepsilon_1, \frac{\varepsilon_I}{\varepsilon_1}, T_{\min}$ are > 0, small, and $T_{\max}$ is large.

By Lemma 2.1, the data of the target $\bar{y}^* = (W_M^*, T^*)$ determines uniquely the target equilibrium x^*, and the target stationary control $u^* = (W_{IF}^*, T_{heat}^*)$. If (P) is satisfied, then $x^* \in Int(\Omega)$, $u^* \in Int(U)$.

Remark 2.2. Notice that this "restricted state space" is in fact very big, and that physically, the state lives in Ω.

Lemma 2.3. *If (P) is satisfied, then the restricted state space is positively invariant under the dynamics (172). On the restricted state space, the system is smooth.*

Exercise 2.2. Prove Lemma 2.3.

For the exercise, use the fact that ΔH in (174) is negative.

2.4.3. Observability

As we said above, the **observed output** is $y = (y_1, y_2) = (T, \frac{1}{\rho})$. The inputs are still $u = (W_{IF}, T_{heat})$. The equations under consideration are (172) and (164), on the compact domain Ω defined above (183). The set of values of control is $U = [\varepsilon_1, 1] \times [T_{\min}, T_{\max}]$. These equations are analytic on $\Omega \times U$, hence, they are analytic on a small open neighborhood $\tilde{\Omega} \times \tilde{U}$ of $\Omega \times U$.

Let us consider the following mapping $\Phi : \tilde{\Omega} \to R^4$,

$$\Phi(T, W_M, W_S, W_I) = \left(T, r = \frac{1}{\rho}, \lambda = \frac{-k_p C_R W_M \Delta H}{M_M C_P}, \mu = \frac{d}{dt}\left(\frac{1}{\rho}\right)\right). \tag{184}$$

Note the important fact that $\frac{d}{dt}(\frac{1}{\rho})$ is well defined, by (182), and does not depend on the controls (W_{IF}, T_{heat}). Hence, Φ is a well-defined **smooth** mapping on Ω, analytic on an open neighborhood $\tilde{\Omega}$ of Ω.

Exercise 2.3. Show that Φ is a diffeomorphism, in restriction to $Int(\Omega)$.

Now, let us write the equations of the reactor in the new coordinates $\Phi(T, W_M, W_S, W_I) = (T, r, \lambda, \mu) = X$.

The important observation is that the equation of T in the new coordinates is

$$\frac{dT}{dt} = \varphi_0(T, r, u_2) + \lambda. \tag{185}$$

Therefore, in the new coordinates, the reactor system becomes

$$(\Sigma_{rea}) \begin{cases} \dot{T} = \varphi_0(T, r, u_2) + \lambda, \\ \dot{r} = \mu, \\ \dot{\lambda} = \varphi_1(X, u), \\ \dot{\mu} = \varphi_2(X, u). \end{cases} \tag{186}$$

The observed outputs being T and r, it is easily shown that the system Σ_{rea} has the following properties:

Lemma 2.4. *Σ_{rea}, defined on the subset $\Omega' = \Phi(Int(\Omega)) \subset R^4$ is observable and uniformly infinitesimally observable.*

Exercise 2.4. Prove the previous Lemma 2.4.

Note: It is clear that the functions $\varphi_0, \varphi_1, \varphi_2$, defined on open subsets of R^3, R^6, R^6, respectively, can be extended as smooth compactly supported functions to all of these vector spaces. Given an arbitrarily large compact subset K in $Int(\Omega)$, this can be done without perturbing them on $\Phi(K) \times U$. In the following, these extensions will be denoted again by $\varphi_0, \varphi_1, \varphi_2$, so that we can assume that the reactor system is globally in the form (186), with compactly supported functions φ_i, $i = 0, 1, 2$.

2.5. Observers

It follows from the **canonical** form (186) of Lemma 2.4, and from the note after, that all of the methods of Chapter 6 can be applied, in order to construct exponential observers for the polymerization reactor, for all U-valued measurable inputs. In fact, again, we are in a case where $d_y = d_u$ but the system Σ_{rea} is observable in all possible strong senses. The theory of Chapter 6 can be applied to this case. We leave it to the reader. **All the types of construction of observers we proposed apply.**

Exercise 2.5. There is a need of a small adaptation for the Luenberger style observer for uniformly infinitesimally observable systems, (Theorem 2.2 of Chapter 6), because we are not in the case $d_y = 1$. Make this adaptation.

2.6. Feedback Stabilization

Using Proposition 2.2, we will be able to construct feedbacks that stabilize asymptotically the reactor within "reasonably large" subdomains of the restricted state space Ω.

Let us consider the following feedbacks:

$$W_{IF}^{\alpha} = \varphi\left(\left(1 + \frac{\rho(W_M, W_S)Vk_d(T)}{Q_{MF}}\right)W_I^* - \alpha(W_I - W_I^*)\right),$$
$$T_{heat}^{\beta} = \psi\left(\left(1 + \frac{Q_{MF}C_P}{U}\right)T^* - \beta(T - T^*) + A\right), \qquad (187)$$

where

$$A = -\frac{Q_{MF}C_{PF}}{U}T_F + \frac{\Delta H\ V}{UM_M}\rho k_p C_R W_M, \qquad (188)$$

where α and β are real numbers, and φ and ψ, smooth, satisfy:

$$\varphi : R \to R,\ \varphi(\theta) = \theta \text{ if } 2\varepsilon_1 \le \theta \le 1 - \varepsilon_1,$$
$$\varphi(\theta) = \varepsilon_1 \text{ if } \theta \le \varepsilon_1,\ \varphi(\theta) = 1 \text{ if } \theta \ge 1,\ \frac{\partial\varphi}{\partial\theta} \ge 0\ \forall\theta,$$
$$\psi : R \to R, \psi(\theta) = \theta \text{ if } T_{\max} - \varepsilon_2 \ge \theta \ge T_{\min} + \varepsilon_2,$$
$$\psi(\theta) = T_{\min} \text{ if } \theta \le T_{\min},\ \psi(\theta) = T_{\max} \text{ if } \theta \ge T_{\max},$$
$$\frac{\partial\psi}{\partial\theta} \ge 0\ \forall\theta,\ \ \varepsilon_2 \text{ small.}$$

For $x \in \Omega$, $W_{IF}^{\alpha}(x) \in [\varepsilon_1, 1]$ and $T_{heat}^{\beta}(x) \in [T_{\min}, T_{\max}]$. Notice also that if $x = x^*$, then $W_{IF}^{\alpha} = W_{IF}^*$, $T_{heat}^{\beta} = T^*$.

Hence, by construction, the equations of the reactor controlled by the feedback just constructed are, in a neighborhood U_{x^*} of the equilibrium x^*:

$$
\begin{aligned}
&(1)\quad \frac{dW_M}{dt} = \frac{Q_{MF}}{\rho V}(W_{MF} - W_M) - k_p W_M C_R,\\
&(2)\quad \frac{dW_S}{dt} = \frac{Q_{MF}}{\rho V}(W_{SF} - W_S) - k_{tts} W_S C_R,\\
&(3)\quad \frac{d(W_I - W_I^*)}{dt} = -\left(k_d + \frac{Q_{MF}}{\rho V}(1+\alpha)\right)(W_I - W_I^*),\\
&(4)\quad \frac{\rho V C_P}{U}\frac{d(T - T^*)}{dt} = -\left(1 + \frac{Q_{MF} C_P}{U} + \beta\right)(T - T^*).
\end{aligned}
\tag{189}
$$

Exercise 2.6. Using Proposition 2.2, show that these equations are (locally) asymptotically stable at the equilibrium x^*, provided that $\alpha, \beta > 0$ or $\alpha, \beta \leq 0$ small.

In fact, there are several other feedbacks that do the same job. In particular, just linear feedbacks would work, but this one is more interesting as we shall show immediately.

It happens that there are "reasonably large" subdomains $\bar{\Omega}$ of Ω that are contained in the basin of attraction of x^* (for (189), the reactor under feedback), and the constraints on the controls are satisfied along the corresponding semitrajectories (i.e., the corresponding feedback controls take values in U). To see this, we have to go again to our special numerical values (174). We will consider the following domain $\bar{\Omega}$:

$$
\bar{\Omega} = \{x | 0 \leq W_M \leq W_{MF}, 0 \leq W_S \leq W_{SF}, \varepsilon_I \leq W_I \leq 0.1, T_0 \leq T \leq 350\},
\tag{190}
$$

where $T_0 \geq T_{\min} + \varepsilon_2$.

Let us recall that our stationary values (x^*, u^*) are

$$
\begin{aligned}
x^* &= (W_M^*, W_S^*, W_I^*, T^*) = (0.7, 0.0999997, 0.03837, 330),\\
u^* &= (W_{IF}^*, T_{heat}^*) = (0.0413, 319.38).
\end{aligned}
\tag{191}
$$

For the domain $\bar{\Omega} \subset \Omega$, there is no additional restriction on W_M, W_S, the maximum of W_I is more than twice the stationary value (and is high, physically), and the temperature T is in a large domain around T^* ($T_{\max}$ is 20 degrees more).

If we show that the feedback controls take values in U on this domain $\bar{\Omega}$, and if these values are such that the functions $\varphi(\theta)$ and$\psi(\theta)$ are just equal to θ, it will follow that $\bar{\Omega}$ is positively invariant for the reactor under feedback.

We already know that Ω is positively invariant, and the equations of the reactor under feedback show that the two last components of the state, W_I and T, cannot go out the bounds just given in the definition (190) of $\bar{\Omega}$.

Let us start with the feedback W_{IF}^0. Clearly, W_{IF}^0 is larger than W_I^* (k_d and ρ are positive functions on Ω, Q_{MF} and V are positive constants). The function k_d is increasing, majorized by $k_d(350.)$, and $\rho \leq \rho_P$ on $\bar{\Omega}$. The numerical values in (174) show that:

$$W_{IF}^0 \leq \left(1 + \frac{kd(350.)\rho_P V}{Q_{MF}}\right) W_I^* = 0.089.$$

We conclude that on $\bar{\Omega}$, $2\varepsilon_1 < W_{IF}^0(x) < 0.1$. The same is true for W_{IF}^α if α is sufficiently small.

Now, $T_{heat}^0 \leq \psi(T^*(1 + \frac{Q_{MF}C_P}{U}))$. On the other hand, the quantity $-A$ in (188) above is positive and can be majorized as follows:

$$-A \leq -A^* = \frac{Q_{MF}C_{PF}}{U} T_F$$
$$- \frac{\Delta H\ V}{U M_M} \rho_P k_p(350.) W_{MF} \sqrt{\rho_P \frac{f}{M_I} W_{\mathrm{Im}ax} \frac{k_d}{k_{tc}}(350.)},$$
$$W_{\mathrm{Im}ax} = 0.1.$$

And T_{heat}^0 is minored by $\psi(A^* + T^*(1 + \frac{Q_{MF}C_P}{U}))$. The quantity $A^* + T^*(1 + \frac{Q_{MF}C_P}{U})$ is computed using the numerical values (174), and is strictly positive. Therefore, for β small enough, T_{heat}^β is between $T_{\min} + \varepsilon_2$ and $T_{\max} - \varepsilon_2$, for $T_{\max}$ large, $T_{\min}, \varepsilon_2 > 0$, small.

We have shown:

Proposition 2.5. *(For the numerical values (174), and for α, β small enough). The domain $\bar{\Omega}$ is positively invariant for the reactor under feedback (189), and the feedback takes values in U on $\bar{\Omega}$.*

Now, one can use again Proposition (2.2) in order to show the following:

Proposition 2.6. *(For the numerical values (174) and for α, β small enough.) The equilibrium x^* is* ***asymptotically stable within*** *$\bar{\Omega}$, for the reactor under feedback (189).*

For other more complicated feedbacks, in the more practical situation of the nonsimplified reactor, see paper [65] and thesis [63].

2.7. Dynamic Output Stabilization.

First, let us notice that we are not in the good situation of Section 3 of Chapter 6. However, we are in the context of Chapter 7, Theorem 1.1 or Theorem 2.3.

Exercise 2.7. Make the necessary adaptations of the proof of Theorem 1.1, Chapter 7 for it applies to the situation of the reactor (we are in the case $d_y \leq d_u$, but $d_y \neq 1$).

The result is that it is possible to stabilize asymptotically the polymerization reactor, using output informations only, at unstable equilibria, within large sets (arbitrarily large compacta in $Int(\bar{\Omega})$).

Remark 2.3. In fact, in practice, we observe that the basin of attraction B of x^* for the reactor under feedback is much larger than $\bar{\Omega}$. We can asymptotically stabilize using the observer within any compact subset of $B \cap Int(\Omega)$.

Appendix

1. Subanalytic Sets

For details on this section, see [5, 6, 25, 45].

1.1. Subanalytic Sets and Subanalytic Mappings

Let M be a real analytic manifold. A subset $S \subset M$ is called **subanalytic** if any point $p \in M$ has an open neighborhood U, such that either $U \cap S$ is empty or

$$U \cap S = \cup_{i=1}^{N}(f_i(A_i)\backslash g_i(B_i)),$$

where $A_1, \ldots, A_N$, $B_1, \ldots, B_N$ are analytic manifolds and $f_i : A_i \to M$, $g_i : B_i \to M$, $1 \leq i \leq N$, are **proper** analytic mappings.

Because the structure of S depends on M, we shall always mention it when we speak about the subanalytic sets. Notation: (S, M). Examples of such sets are: (i) the **closed** analytic subsets of M, (ii) complements of such sets, (iii) polyhedra in Euclidean spaces.

Subanalytic mappings: Let (S_i, M_i), $i = 1, 2$, be two subanalytic sets and $f : S_1 \to S_2$ a **continuous mapping**. We say that f is subanalytic if its graph $G(f)$ considered as a subset of $M_1 \times M_2$ is subanalytic in $M_1 \times M_2$.

1.2. Stability Properties of Subanalytic Sets

M, N will denote analytic manifolds. $Sub(M)$, (resp. $Sub(N)$) will denote the family of all subanalytic subsets of M (resp. N). Then

1. If $S, T \in Sub(M)$; $S \cup T \in Sub(M)$, $S \cap T \in Sub(M)$, $S\backslash T \in Sub(M)$. If $S \in Sub(M)$, $T \in Sub(N)$, $S \times T \in Sub(M \times N)$,
2. If $S \in Sub(M)$ and $\bar{S}$ is the topological closure of S in M, $\bar{S} \in Sub(M)$; from 1, it follows that the frontier $\bar{S}\backslash S$ of S in M is in $Sub(M)$,
3. Each connected component of an $S \in Sub(M)$ belongs to $Sub(M)$. The family of the connected components of S is locally finite (i.e., any

point $p \in M$ has a neighborhood meeting only a finite number of these components),

4. If $S \in Sub(M)$ and $f : S \to N$ is subanalytic ($N \in Sub(N)$!), then, **if** f **is proper**, $f(S) \in Sub(N)$,
5. If f is proper, $S \in Sub(N)$, then $f^{-1}(S) \in Sub(M)$.

1.3. Singular Points of Subanalytic Sets, Dimension

Let M be an analytic manifold, of dimension m. Given $S \in Sub(M)$, a point p is called **regular** if it has an open neighborhood O such that $O \cap S$ is a closed analytic submanifold of O. We denote by Reg(S) the set of all regular points of S, and by Sing(S) its complement in S. The points in Sing(S) are called **singular**.

Reg(S) is open in S and is an analytic submanifold of M. Its dimension (i.e., the maximum dimension of its connected components) will be called the **dimension of** S. It coincides with the Hurwicz–Wallman dimension of S as a topological space.

Proposition 1.1. (*Theorem 7.2 p. 37, [5]*) *If* $S \in Sub(M)$, *Sing*(S) *is a subanalytic set in* M *and is closed in* S. *Reg*(S) *is also subanalytic in* M. dim(*Sing*(S)) $<$ dim(S) *always.* (dim($\emptyset$) $= -1$).

1.4. Whitney Stratifications

Let M be an analytic manifold, countable at infinity. Let $S \in Sub(M)$.

Theorem 1.2. *There exists a partition* $\mathcal{P}$ *of* S *such that:*

(i) Each $P \in \mathcal{P}$ *is a real analytic connected submanifold in* M, *and is subanalytic in* M,

(ii) $\mathcal{P}$ *is locally finite in* M *(i.e., if* $p \in M$, *there exits a neighborhood* U *of* p *such that* $\{A \in \mathcal{P} \mid A \cap U \neq \emptyset\}$ *is finite,*

(iii) For any $A, B \in \mathcal{P}$ *such that* $\bar{A} \cap B$ *is nonempty,* B *is in fact contained in* $\bar{A}$,

(iv) For any $A, B \in \mathcal{P}$ *such that* $\bar{A} \supset B$, (A, B) *is a Whitney pair.*

Whitney pair (X, Y) **at** $y \in Y$: let X, Y be connected submanifolds in M, and y a point in Y such that $y \in \bar{X}$. We say that (X, Y) is a Whitney pair at y if there exists a chart (Ω, ψ) of M at y having the following property:

- Given any sequences (x_n), (y_n) such that (α) $(x_n | n \in \mathcal{N}) \in \Omega \cap X$, $(y_n | n \in \mathcal{N}) \in \Omega \cap Y$, $x_n \neq y_n$ for all $n \in \mathcal{N}$, $x_n \to y$, $y_n \to y$ as

$n \to +\infty$, (β) the lines $R(\psi(y_n) - \psi(x_n))$ converge to a line l, (γ) the vector spaces $(T\psi(T_{x_n}X); n \in \mathcal{N})$ converge to a vector subspace T of R^m (the convergence is in the Grassmannian $G(m, p)$, $p = \dim(X)$, $m = \dim(M)$);

then, $T \supset l$.

A couple (X, Y) is called a Whitney pair if $\bar{X} \supset Y$ and (X, Y) is a Whitney pair at all points of Y.

Let M, N be analytic manifolds both countable at infinity. Let $S \in Sub(M)$, $T \in Sub(N)$, and $f : S \to T$ be a subanalytic mapping. We assume that f is **proper**.

Theorem 1.3. *There exist partitions $\mathcal{P}_S, \mathcal{P}_T$ of S and T, respectively, satisfying the conditions (i), (ii), (iii), and (iv) of Theorem 1.2, and such that for any $X \in \mathcal{P}_S$, there exists a $Y \in \mathcal{P}_T$ containing $f(X)$, and the restriction $f_{|X}$ is a C^ω submersion.*

1.5. Conditions a and b

The property of a couple (X, Y) at y in the definition of a Whitney pair has been called by Whitney: **condition *b***. He has also introduced a weaker condition, **condition *a*** as follows:

- Given a pair (X, Y) of connected C^1 manifolds and a point $y \in Y \cap \bar{X}$, we say that the pair (X, Y) satisfies the condition a at y if there exists a chart (Ω, ψ) of M at y with the following property: Given any sequence $(x_n | n \in \mathcal{N})$ such that (α) $(x_n | n \in \mathcal{N}) \subset \Omega \cap X$, $x_n \neq y$ for all $n \in \mathcal{N}$, and $x_n \to y$ an $n \to +\infty$, (β) the vector spaces $\{T\psi(T_{x_n}X) \mid n \in \mathcal{N}\}$ converge to a vector subspace T of R^m, then $T \supset T\psi(T_y Y)$.

It is known and easy to see that condition b implies condition a (but not vice versa). It is also easy to see that these conditions are independent of the chart (Ω, ψ): if one is true for a chart, it is true for all the others. In fact, at the cost of introducing additional concepts, one can state conditions a and b independently of the choice of any chart.

2. Transversality

2.1. The Concept of Transversality

Let $r \in \mathcal{N}$ or $r = \infty$ or $r = \omega$. Let M, P be two C^r manifolds, and N be a C^r submanifold of P. Except explicit mention of the contrary, we allow both M and N to have corners. Denote by $f : M \to P$ a C^r mapping.

Definition 2.1.

(i) let $x \in M$. Then f is said to be transversal to N at x if either $f(x) \notin N$ or $f(x) \in N$ and

$$T_x f(T_x M) + T_{f(x)}(N) = T_{f(x)}P.$$

(All these concepts have a meaning at any point of a manifold with corners.)

If $S \subset M$ is a subset, we say that f is transversal to N on S if it is transversal to N at all $x \in S$.

(ii) If M is a submanifold of P and $f : M \to P$ is the canonical injection of M into P, we say that M and N are transversal at x (resp. on S).

Remark 2.1. It is clear that the transversality relation defined in (ii) is symmetric in M and N.

2.2. *Transversality Theorems*

2.2.1. *Stability*

If $r = \omega$, the C^r topology is the C^∞ topology.

Proposition 2.1. *If $x_0 \in M$, $f(x_0) \in N$ and f is transversal to N at x_0, then there exists a neighborhood O of f in $C^r(M, P)$ endowed with the C^r topology (if $r = \omega$, the C^∞ topology), and a neighborhood U of x_0 in M such that for any $g \in O$, any $x \in U$, g is transversal to N at x.*

Remark 2.2. Proposition 2.1 is still true if we do not assume that $f(x_0) \in N$, provided that N is closed (in P). Otherwise, it is false as the following example shows.

Example 2.1. Let $r = \infty$, $M = R$, $P = R^2$ with canonical coordinates (x, y). Let $N = \{(x, 0) \mid x > 0\}$, $f : R \to R^2$, $f(t) = (t, t^2)$. The mapping f is transversal to N at zero, because $f(0) = (0, 0) \notin N$. But for any neighborhood O of f in $C^\infty(M, P)$, any neighborhood U of 0 in $M = R$, there exists an $\varepsilon \in U$ and an $f_\varepsilon \in O$, where $f_\varepsilon : M \to P$ is the mapping $f_\varepsilon(t) = (t, (t - \varepsilon)^2)$, and f_ε is not transversal to N at ε.

2.2.2. *Abstract Transversality Theorem*

Here, we assume that M is without corners, and that M and P are second countable. $r \in \mathcal{N} \cup \{+\infty\}$. Let T be a Hausdorff topological space and

$\rho : T \to C^r(M, P)$ be a continuous mapping for either the C^r or the Whitney C^r topology, having the following property:

- For each $t \in T$, there exists a C^r manifold A, a point $a \in A$, and a continuous mapping $\varphi : A \to T$ such that (i) $\varphi(a) = t$, and (ii) the mapping $\Phi : A \times M \to P$, $\Phi(a, x) = (\rho \circ \varphi(a))[x]$ (evaluation at x !) is C^r and transversal to N.

Theorem 2.2. *(Abstract transversality theorem)*
If $r >$ dim$(M)-$ *Codim*(N, P), *then,*

(i) the set T_r of all $t \in T$ such that $\rho(t)$ is transversal to N is a dense residual subset in T,
(ii) if N is closed and either ρ is continuous in the Whitney C^r topology on $C^r(M, P)$, or M is compact, then T_r is open dense in T.
(iii) if Codim$(N, P) >$ dim(M), *the result of (ii) is still true if we take* $r = 0$. *Transversal means that the mapping $f : M \to P$* ***avoids*** *N.*

2.2.3. *Transversality to Stratified Sets*

M is again without corners. Assume now that N is a C^r stratified subset of P with stratification $\mathfrak{N}$ (we assume that the stratifications are locally finite !).

Definition 2.2. f is said to be transversal to the stratification $\mathfrak{N}$ if for any $S \in \mathfrak{N}$, f is transversal to S (each stratum $S \in \mathfrak{N}$ is a C^r submanifold of P).

In general, the condition for a map to be transversal to a stratified set is not open, even if the set is closed. To ensure this fact, we have to consider special stratifications.

Definition 2.3. A stratification $\mathfrak{N}$ of a subset N of P is called a "Whitney-a" stratification if (i) for each stratum $S \in \mathfrak{N}$, $\bar{S}$ is a union of strata, and if S_1, $S_2 \in \mathfrak{N}$, $S_2 \cap \bar{S}_1 \neq \emptyset$, then $S_2 \subset \bar{S}_1$, (ii) for each pair S_1, $S_2 \in \mathfrak{N}$ such that $S_2 \subset \bar{S}_1$, the pair (S_1, S_2) satisfies the condition "Whitney-a" at each point of S_2.

Let us assume that M and P are second countable.

Proposition 2.3. *Let the stratification $\mathfrak{N}$ of N be a C^r Whitney-a stratification. Assume that N is closed. Then, the subset of all $f \in C^r(M, P)$, which are transversal to $\mathfrak{N}$, is open in the C^r Whitney topology.*

2.2.4. Transversality on Corner Manifolds

Let $\Pi : E \to B$ be a C^∞ vector bundle over the C^∞ manifold B. Let $J^k(\Pi)$ be the bundle of k-jets of sections of Π ($k \in \mathcal{N}^*$). Let M be a C^∞ submanifold **with corners** of B, **of the same dimension**. Denote by $\Gamma^\infty(\Pi)$ (resp. $\Gamma^\infty(\Pi_M)$) the space of all C^∞ sections of Π (respectively of Π_M, restriction of Π to $\Pi^{-1}(M)$). Let $N \subset J^k(\Pi)$ be a closed C^∞ stratified subset of $J^k(\Pi)$, endowed with a Whitney-a stratification $\mathfrak{N}$. We assume that B is second countable, and we will restrict it in the following tacitly to appropriate neighborhoods of M.

Theorem 2.4. *The subset of all $f \in \Gamma^\infty(\Pi_M)$ such that the mapping $j^k f : M \to J^k(\Pi)$ is transversal to $\mathfrak{N}$ is open, dense in the C^∞ Whitney topology on $\Gamma^\infty(\Pi_M)$.*

Proof. That the subset is open is a consequence of Proposition 2.3 (actually an easy modification of it).

To see that it is dense, we use Stein extension theorem to construct an extension operator $E : \Gamma^\infty(\Pi_M) \to \Gamma^\infty(\Pi)$, which is continuous for the Whitney C^∞ topologies on $\Gamma^\infty(\Pi_M)$ and $\Gamma^\infty(\Pi)$. Let $\mathcal{T}_r^M$ (resp. $\mathcal{T}_r$) be the set of all sections $f \in \Gamma^\infty(\Pi_M)$ (resp. $\Gamma^\infty(\Pi)$) such that $j^k f$ is transversal to N. Then, $res(\mathcal{T}_r) \subset \mathcal{T}_r^M$, where $res : \Gamma^\infty(\Pi) \to \Gamma^\infty(\Pi_M)$ is the restriction operator. If $f \in \Gamma^\infty(\Pi_M)$, then any neighborhood of $E(f)$ contains elements of $\mathcal{T}_r$. Now let U be a neighborhood of f in $\Gamma^\infty(\Pi_M)$, V its inverse image, $V = res^{-1}(U)$. Then, V is a neighborhood of $E(f)$. It contains an element g of $\mathcal{T}_r$. Hence, $res(g) \in \mathcal{T}_r^M \cap U$. ■

2.3. The Precise Arguments We Use in Chapter 4

All the arguments we use are consequences of the results just stated. First, X is a smooth manifold without corners, and $U = I^{d_u}$, where I is a closed bounded interval. The set of systems is the set of (C^r, $r = 1, \ldots \infty$) couples (f, h), where f is a U-parametrized vector field over X, and $h : X \times U \to R^{d_y}$, or $h : X \to R^{d_y}$. The set $J^k S$ of k-jets of elements of S is a vector bundle over $X \times U$. If X is compact, S can be given the structure of a Banach space. If X is not compact, we endow it with the (C^r or C^∞) Whitney topology. We assume that r is sufficiently large.

Theorem 2.5. *N is a C^r closed stratified subset of $J^k S$, with Whitney-a stratification $\mathfrak{N}$. The subset of elements $\Sigma = (f, h)$ of S such that $j^k \Sigma$ is transversal to $\mathfrak{N}$ is open dense.*

Now, $J^k S_*^2$ denotes the restriction of $(J^k S)^2$ to $Z = ((X \times X)\backslash \Delta X) \times \Delta U$. For $\Sigma \in S$, $(j^k \Sigma)^2 : Z \to J^k S_*^2$ is the map $(x_1, x_2, u) \to (j^k \Sigma(x_1, u), j^k \Sigma(x_2, u))$. We chose W, a compact subset of $((X \times X)\backslash \Delta X)$.

Theorem 2.6. *N is a C^r closed stratified subset of $J^k S_*^2$, with Whitney-a stratification $\mathfrak{N}$. The set of $\Sigma \in S$, such that $(j^k \Sigma)^2$ is transversal to $\mathfrak{N}$ on $W \times \Delta U$, is open dense.*

These two theorems contain what is needed for the proof of Theorems 2.1, 2.2, and 2.3 of Chapter 4. The considerations in this appendix show that these theorems are valid for X noncompact and for the Whitney topology.

For a more detailed and elementary exposition of what is needed if X is compact (and U without corners), the reader can consult [1], as well as [20] for transversality to stratified sets.

For the proof of Theorem 2.4 of Chapter 4, which is one of the two main steps in proving the final approximation theorem (Theorem 5.1) in the same chapter, the following theorem is needed.

X is **analytic, compact.** S is now a set of systems of the form S_K, constructed in Chapter 4, Section 1. In that case, S_K has also the structure of a Banach space.

Theorem 2.7. *Theorems 2.5 and 2.6 are still valid for S_K.*

3. Lyapunov's Theory

For details on this section, see [35–37, 40].

3.1. The Concept of Stability and Asymptotic Stability

Let M be a C^∞ Manifold, F a C^∞ vector field on M, and $x_0 \in M$ a stationary point of F. For $x \in M$, the maximal positive semitrajectory of x for F is a C^∞ curve:

$$t \in [0, e(x)[\to \varphi_t(x) \in M,$$

where $e(x) > 0$ and possibly $e(x) = +\infty$, such that:

(i) $\frac{d\varphi_t(x)}{dt} = F(\varphi_t(x))$ for all $t \in [0, e(x)[$,
(ii) $\varphi_0(x) = x$,
(iii) $t \to \varphi_t(x)$ is maximal, that is, either $e(x) = +\infty$, or $e(x) < +\infty$ and the set

$$\cap_{0<r<e(x)} \overline{\{\varphi_t(x) \mid r \leq t < e(x)\}},$$

is empty, where the bar denotes the closure of the set in M. (It is equivalent to say that the trajectory $t \to \varphi_t(x)$ cannot be extended beyond $e(x)$.) The maximal positive trajectory of a $x \in M$ is **unique**.

Definition 3.1. (Stability). x_0 is called stable for F if any neighborhood N_0 of x_0 in M contains a subneighborhood N_1 of x_0 such that the maximal positive semitrajectory of any $x \in N_1$ is contained in N_0 ($\varphi_t(x) \in N_0$ for all $t \in [0, e(x)[$).

3.1.1. Consequences of Stability

1. If x_0 is stable, then it has a neighborhood U such that $e(x) = +\infty$ for all $x \in U$. To see this, choose any compact neighborhood N_0 of x_0. Then, we can take $U = N_1$ associated to N_0 by the definition. In fact, because N_0 is compact and $\varphi_t(x) \in N_0$ for all $t \in [0, e(x)[$, $\bigcap_{0<r<e(x)} \overline{\{\varphi_t(x) \mid r \leq t < e(x)\}}$ cannot be empty (decreasing intersection of compact nonempty sets).
2. Any neighborhood V of x_0 in M contains a compact neighborhood K of x_0 such that the maximal positive semitrajectory of any $x \in K$ is contained in K : choose a compact neighborhood N_0 of x_0 contained in V. Let K be the set of all points in N_0 whose maximal positive semitrajectory is contained in N_0. Then (a) K is a neighborhood of x_0 because it contains N_1; (b) $e(x) = +\infty$ for all $x \in K$ by what we have seen above; (c) $\varphi_t(K) \subset K$ for all $t \geq 0$: for x and for $t > 0$, the positive semitrajectory of $\varphi_t(x)$ is contained in that of x; (d) K is closed and hence compact: let $\bar{x}$ be in the closure of K. Then $\bar{x} \in N_0$. If $\bar{x} \notin K$, then there exists a $\bar{t}$ such that $\varphi_{\bar{t}}(\bar{x}) \in M \backslash N_0$. Because $M \backslash N_0$ is open, there exists a neighborhood U of $\bar{x}$ such that $e(x) > \bar{t}$ for all $x \in U$ and $\varphi_{\bar{t}}(U) \subset M \backslash N_0$. Now, $U \cap K$ is not empty. So we have a contradiction because $\varphi_t(U \cap K) \subset N_0$ for all $t \geq 0$.

We call a compact subset K of V having these properties (a), (b), (c), and (d) with respect to V, "adapted to V."

Definition 3.2. (Asymptotic stability). We say that x_0 is asymptotically stable for F if it has a neighborhood U such that:

(i) $e(x) = +\infty$ for all $x \in U$,
(ii) $\varphi_t(U) \subset U$ for all $t \geq 0$,
(iii) $\bigcap_{t \geq 0} \varphi_t(U) = \{x_0\}$.

There is a simple criterion to see if x_0 is asymptotically stable:

Lemma 3.1. *x_0 is asymptotically stable if and only if:*

(i) x_0 is stable,

(ii) there exists a neighborhood V of x_0 such that $\varphi_t(x)$ tends to x_0 when t tends to $+\infty$, for all $x \in V$ (we have seen in (1) above that x_0 has a neighborhood N such that $e(x) = +\infty$ for all $x \in N$).

Proof. Assume that x_0 satisfies the conditions (i) and (ii) of the lemma. Choose a $K \subset V$ adapted to V. All we have to do to prove that x_0 is asymptotically stable is to show that $\cap_{t\geq 0}\varphi_t(K) = \{x_0\}$. Were it otherwise, there would exist an open neighborhood O of x_0 such that $(M\backslash O) \cap_{t\geq 0} \varphi_t(K)$ would not be empty. Choose a neighborhood L of x_0 adapted to O.

The interior L° of L is a neighborhood of x_0 and for any $x \in K$, there exists a $t_x > 0$ such that $\varphi_{t_x}(x) \subset L^\circ$. Then, there exists a neighborhood P_x of x such that $e(x') > t_x$ for all $x' \in P_x$ and $\varphi_{t_x}(P_x) \subset L^\circ$. Choose a finite set $\{x_1, \ldots, x_n\} \subset K$ such that $\cup_{i=1}^n P_{x_i} \supset K$. Take $\bar{t} = \max_{1\leq i\leq n} t_{n_i}$. Because $\varphi_t(L) \subset L$ for all $t \geq 0$, $\varphi_{\bar{t}}(P_{x_i}) \subset L$ for all i, $1 \leq i \leq n$. Then, $\varphi_{\bar{t}}(K) \subset L \subset O$. Hence, $\cap_{t\geq 0}\varphi_t(K) \cap (M\backslash O)$ is empty, which is a contradiction.

Assume now that x_0 is asymptotically stable. We prove that each relatively compact open neighborhood $V \subset \bar{V} \subset U$ of x_0 contains an open neighborhood Ω of x_0 such that $\cup_{t\geq 0}\varphi_t(\Omega) \subset V$. In fact, define the set Ω as follows: $\Omega = \{x \in V \mid \varphi_t(x) \in V \text{ for all } t \geq 0\}$. Clearly, $x_0 \in \Omega$. We have to show that it is open or that $V\backslash\Omega$ is closed (in V !). $V\backslash\Omega = \{x \mid \exists\, t'_x > 0, \varphi_{t'_x}(x) \in M\backslash V\}$. For each $x \in V\backslash\Omega$, let $\tau(x) = \inf\{t \mid \varphi_t(x) \notin V\}$. Hence, $\varphi_{\tau(x)}(x) \in \bar{V}\backslash V$ for all $x \in V\backslash\Omega$. Now, let x_n be a sequence in $V\backslash\Omega$ converging to x_∞ (in V !). We have to show that $x_\infty \in V\backslash\Omega$. Because the $\varphi_{\tau(x)}(x) \in \bar{V}\backslash V$, which is compact, choosing a subsequence, we can assume that $\varphi_{\tau(x_n)}(x_n) \to \xi$ as $n \to +\infty$ and that $\{\tau(x_n) \mid n \in \mathcal{N}\}$ converges either to a finite number τ_∞ or to $+\infty$. $\xi \in \bar{V}\backslash V$. If $\tau(x_n) \to \tau_\infty$, then $\xi = \varphi_{\tau_\infty}(x_\infty)$ and $x_\infty \in V\backslash\Omega$. If $\tau(x_n) \to +\infty$, then $\varphi_{-t}(\xi)$ is defined for all $t \geq 0$ and the negative semitrajectory γ of ξ, $\gamma = \{\varphi_{-t}(\xi) | t \geq 0\}$ is contained in $\bar{V}$ as we show below. Then, for any $t \geq 0$, $\varphi_t(\bar{V}) \supset \varphi_t(\gamma) \supset \gamma$. So, $\cap_{t\geq 0}\varphi_t(U) \supset \cap_{t\geq 0}\varphi_t(\bar{V}) \supset \gamma \neq \{x_0\}$, which is a contradiction.

Hence, Ω is open. Clearly, $\varphi_t(\Omega) \subset \Omega$, $\varphi_t(\bar{\Omega}) \subset \bar{\Omega}$ for all $t \geq 0$. Because $\bar{\Omega}$ is compact, it is easy to see that $\varphi_t(x) \to x_0$ as $t \to +\infty$ for any $x \in \bar{\Omega}$: let N be any open neighborhood of x_0. Then, there is a $T > 0$ such that $\varphi_T(\bar{\Omega}) \subset N$. Otherwise, $\varphi_t(\bar{\Omega}) \cap (M\backslash N)$ is nonempty for all $t > 0$. Hence, $(\cap_{t\geq 0}\varphi_t(\bar{\Omega})) \cap (M\backslash N)$ is nonempty, which is a contradiction.

To see that $\varphi_{-t}(\xi)$ is defined for all $t \geq 0$, and that $\gamma \subset \bar{V}$, assume these assertions are not true. Then there exists a $\bar{t} > 0$ such that $\varphi_{-s}(\xi)$

is defined for all $0 \le s \le \bar{t}$ and $\varphi_{-\bar{t}}(\xi) \notin \bar{V}$. Because $M \backslash \bar{V}$ is open, there exists a neighborhood Λ of ξ such that φ_{-s} is defined on Λ for all $0 \le s \le \bar{t}$, $\varphi_{-\bar{t}}(\Lambda) \subset M \backslash \bar{V}$. For $n > \bar{n}$, $\tau(x_n) > \bar{t}$ and $\varphi_{\tau(x_n)}(x_n) \in \Lambda$. Hence, $\varphi_{\tau(x_n)-\bar{t}}(x_n) \in M \backslash \bar{V}$. However, by the definition of $\tau(x_n)$, $\varphi_{\tau(x_n)-\bar{t}}(x_n) \in V$, since $0 \le \tau(x_n) - \bar{t} < \tau(x_n)$. This is a contradiction. ■

3.2. Lyapunov's Functions

Definition 3.3. A function $V : O \to R$ defined in an open neighborhood O of x_0 is called a Lyapunov function for F if:

(i) V is continuous, $V(x_0) = 0$ and $V(x) > 0$ if $x \in O \backslash \{x_0\}$.
(ii) For any $x \in O$, the Lie derivative $\dot{V}(x)$ of V in the direction of F exists, is continuous as a function of x, and $\dot{V}(x) \le 0$ for all $x \in O$.

Theorem 3.2. (Stability). *If F has a Lyapunov function $V : O \to R$ at x_0, then x_0 is stable.*

Proof. Choose a coordinate system $(U, x_1, \ldots, x_m)$ of M at x_0 such that: (i) $x_i(x_0) = 0$, $1 \le i \le m$; (ii) there exists a $r_0 > 0$ such that the image of U by $x_1 \times \ldots \times x_m : U \to R^m$ is the ball $B^m(0, 2r_0)$; and (iii) $U \subset O$.

For any $r \in [0, r_0[$, denote by $S(r)$ (resp. $B(r)$) the inverse images of the sphere $S(0,r)$ (resp. the open ball $B(0,r)$) by $x_1 \times \ldots \times x_m$. Hence, $S(0,r) = \{q \in U \mid \sum_{k=1}^m x_k(q)^2 = r^2\}$, $B(0,r) = \{q \in U \mid \sum_{k=1}^m x_k(q)^2 < r^2\}$. Finally, let $\bar{B}(r) = B(r) \cup S(r) = \{q \in U \mid \sum_{k=1}^m x_k(q)^2 \le r^2\}$. The sets $S(r)$, $\bar{B}(r)$ are compact subsets of U for $r \in [0, r_0[$.

Define $\mu : [0, r_0] \to R_+^* =]0, +\infty[$ as follows:

$$\mu(r) = \inf\{V(x) \mid r \le \|x\| < r_0\}.$$

The sets $\bar{B}(r)$, for $r \in]0, r_0]$ form a neighborhood basis of x_0 in M. To prove stability of x_0 it is sufficient to show that for any $r \in]0, r_0]$, there exists a neighborhood K_r of x_0, $K_r \subset \bar{B}(r)$ such that the maximal positive trajectory of F (in M !) starting from $x \in K_r$ is contained in $\bar{B}(r)$.

Take $K_r = \bar{B}(r) \cap \{x \mid x \in O,\ V(x) \le \frac{\mu(r)}{2}\}$. Because V is continuous, K_r is closed ($O \supset U \supset \bar{B}(r)$!). Also, $K_r \cap S(r) = \emptyset$ because $V(x) \ge \mu(r)$ for all $x \in S(r)$. Hence, $K_r \subset B(r)$. Let $x \in K_r$ and let $\xi_x : [0, e(x)[\to M$ be the positive maximal semitrajectory starting at x. Then, $\xi_x^{-1}(B(r))$ is not empty. Let $[0, \tau[$ be its connected component that contains 0. The function $t \in [0, \tau[\to V(\xi_x(t))$ is nonincreasing because $\frac{d}{dt} V(\xi_x(t)) \le 0$. Hence, $V(\xi_x(t)) \le V(\xi_x(0)) = V(x) \le \frac{\mu(r)}{2}$ for all $t \in [0, \tau[$. If $\tau < e(x)$, we get $V(\xi_x(\tau)) \le \frac{\mu(r)}{2}$ and because $\xi_x(\tau) \in \bar{B}(r)$, $\xi_x(\tau) \in K_r \subset B(r)$, which contradicts the definition of τ. Hence, $\tau = e(x)$. But, then, the maximal positive semitrajectory ξ_x lies in $\bar{B}(r)$, and hence, $e(x) = +\infty$. In fact, ξ_x lies in K_r. This shows the stability of x_0. ∎

Theorem 3.3. (Asymptotic stability). *If F has a Lyapunov function $V : O \to R$ at x_0 such that $\dot{V}(x) < 0$ for all $x \in O\backslash\{x_0\}$, then x_0 is asymptotically stable.*

Proof. We use the same considerations and notations as in the proof of Theorem 3.2. By Lemma 3.1, all we have to do is to show that there exists a neighborhood L of x_0 in M such that $\xi_x(t) \to x_0$ as $t \to e(x)$ for any $x \in L$ (it implies the fact that $e(x) = +\infty$!). As L, we chose $K_{\frac{r_0}{2}}$. We have seen in the proof of Theorem 3.2 that if $x \in K_{\frac{r_0}{2}}$ its maximal positive semitrajectory $\xi_x : [0, e(x)[\to M$ lies in $K_{\frac{r_0}{2}}$ and $e(x) = +\infty$. Assume that a sequence $(\xi_x(t_n) \mid n \in \mathcal{N})$, where $t_{n \to +\infty}$ as $n \to +\infty$, converges to a point $y \neq x_0$. Because $K_{\frac{r_0}{2}}$ is closed, $y \in K_{\frac{r_0}{2}}$. Because $y \neq x_0$, $\dot{V}(y) < 0$. Hence, there exists $h > 0$, $\varepsilon > 0$ such that $V(\xi_y(h)) < V(y) - \varepsilon$. By taking a subsequence, we can assume that $\xi_x(t_n) \to y$ as $n \to +\infty$ and $t_{n+1} - t_n \geq h$ for all $n \in \mathcal{N}$. Then, $\xi_x(t_n + h) \to \xi_y(h)$ as $n \to +\infty$. Hence, $\lim_{n\to+\infty} V(\xi_x(t_n + h)) \geq V(\xi_x(t_{n+1}))$ for $n \in \mathcal{N}$. Hence, $\lim_{n\to+\infty} V(\xi_x(t_n + h)) \geq \lim_{n\to+\infty} V(\xi_x(t_{n+1})) = V(y)$. We get a contradiction. ∎

3.3. *Inverse Lyapunov's Theorems and Construction of Lyapunov's Functions*

Here, we assume M to be paracompact. Hence, it is a metric space. Denote by d a distance function on M (for instance, defined by a Riemannian metric on M).

Assume that x_0 is asymptotically stable. Let A denote the basin of attraction of x_0, that is, the set of all $x \in M$ such that the maximal positive semitrajectory $\varphi_t(x)$ of F starting at x at $t = 0$ is defined on $[0, +\infty[$ and $\varphi_t(x)$ tends to 0 as $t \to +\infty$. A is **an open set.**

Theorem 3.4. *There exists a C^∞ function $V : A \to R^+$ having the following properties:*

(i) *$V(x) > 0$ for all $x \in A\backslash\{x_0\}$, $V(x_0) = 0$,*

(ii) *There exists continuous functions $a, b : R_+ \to R$ such that $a(0) = b(0) = 0$, $a(t), b(t) > 0$ for $t \in R_+^*$ such that, for all $x \in A$:*

$$dV(F)(x) \leq -a(d(x, x_0))b(V(x)),$$

(iii) V is proper.

Hence, V is a Lyapunov's function for x_0 in a very strong sense.

For the proof, see [35, 40].

3.4. Lasalle's Theorem

Lasalle's theorem is a generalization of the Lyapunov's direct method.

Theorem 3.5. *Let $V : M \to R^+$ be a continuous function that is proper, such that the Lie derivative of V with respect to F exists at each point of M. Denote it by $\dot{V}$ and assume that $\dot{V}$ is continuous and ≤ 0 everywhere.*

Then, for any $x \in M$, the maximal positive semitrajectory $\varphi_t(x)$ of x for F is defined on $[0, +\infty[$, and its ω-limit set lies in $\{z \in M \mid \dot{V}(z) = 0, V(z) = \inf_{t \geq 0} V(\varphi_t(x))\}$.

Proof. $t \to V(\varphi_t(x))$ is a decreasing function of t. Hence, $\lim_{t \to +\infty} V(\varphi_t(x)) = \inf_{t \geq 0} V(\varphi_t(x)) = a$, say.

Let $y \in \omega(x)$. Then, $y = \lim_{n \to +\infty} \varphi_{t_n}(x)$ where $t_n \to +\infty$ as $n \to +\infty$. Then, $V(y) = a$. If $\dot{V}(y) \neq 0$, $\dot{V}(y) < 0$ and $F(y) \neq 0$. Using the flow box theorem, we see that for $\bar{t}$ big enough $V(\varphi_{\bar{t}}(x)) < a$. But $V(\varphi_{\bar{t}}(x)) \geq V(y) = a$, which is a contradiction. ∎

The same reasoning when V is not proper and not necessarily ≥ 0 shows that the following proposition holds.

Proposition 3.6. *Assume that V is as in Theorem 3.5* **but V is not proper.** *Then, for all $x_0 \in M$ such that $x(t, x_0)$ is defined for all $t \geq 0$, the ω-limit set $\omega(x_0)$ is empty or is contained in the set $\{z \in M \mid \dot{V}(z) = 0\}$.*

4. Center Manifold Theory

4.1. Definitions and Center Manifold Theorem

Let M be an open subset in R^m, F a C^∞ vector field on M, and $x_0 \in M$ a stationary point of F. The vector field F has a linear part at x_0, which is a linear mapping $LF : R^m \to R^m$, whose matrix in the canonical coordinates $(x_1, \ldots, x_m)$ is $\{\frac{\partial F_i}{\partial x_j}(x_0) \mid 1 \leq i, j \leq m\}$, where $F_1, \ldots, F_m$ are the components of F.

The linear part LF is an important invariant of F at x_0. Hence, it must have a coordinate-free interpretation. Let $]-\varepsilon,\varepsilon[\times U \to M$, $(t,x)\to \varphi_t(x)$ be the local flow of F, for U a neighborhood of x_0.

Let $\Phi_t : T_{x_0}R^m \to T_{x_0}R^m$, $t\in]-\varepsilon,\varepsilon[$ be the tangent mapping of φ_t at x_0. Then, $LF = \frac{d\Phi_t}{dt}|_{t=0}$.

Assume now that LF satisfies the following conditions:

R^m has a direct sum decomposition $R^m = n\oplus h$, invariant under LF (i.e., $LF(n)\subset n$, $LF(h)\subset h$), such that the restriction of LF to n has only purely imaginary eigenvalues, and its restriction to h has no purely imaginary eigenvalue (Note: 0 is purely imaginary, $0=\sqrt{-1}.0$).

Theorem 4.1. (Center manifold theorem). *Under the above assumptions, for any $r\in\mathcal{N}^*$, there exists an open neighborhood U of x_0, a C^r closed submanifold N of U such that: (i) $x_0\in N$ and $T_{x_0}N = n$, and (ii) for any $x\in N$, the maximum trajectory of F in U passing through x at time 0 is contained in N.*

Such a closed submanifold N of U also satisfies the following property:

For any $x\in U$ such that the maximal positive (resp. negative) semitrajectory of F in U starting (resp. ending) for $t=0$ at x, is defined for all $t\geq 0$ (resp. $t\leq 0$), then the set $\omega_U(x)$(resp. $\alpha_U(x)$) of all limit points of that semitrajectory as $t\to+\infty$ (rep. $t\to-\infty$) is contained in N.

Definition 4.1. A couple (U, N) satisfying the conditions of Theorem 4.1 is called a **local center manifold** of class C^r of F, at x_0.

Let us make two important remarks.

Remark 4.1. In general, there does not exist a local center manifold of class C^∞ of F at x_0.

Example 4.1. (Van der Strien). $M = R^3$ with canonical coordinates (x, y, z), $x_0 = 0$, and F is the field $(-yx - x^3)\frac{\partial}{\partial x} + (-z + x^2)\frac{\partial}{\partial z}$. Then, h is the z-axis, n is the plane $\{z=0\}$. Assume that F has a local center manifold of class C^∞ at x_0, (U, N). Because $T_{x_0}N = \{z=0\}$, restricting U and N, we can assume that $U = O\times]-\varepsilon,\varepsilon[$, where $O = \{(x,y,0)\mid x^2+y^2<r^2\}$, $r,\varepsilon>0$, and N is the graph of a C^∞ function $\nu : O\to]-\varepsilon,\varepsilon[$.

Condition (ii) of Theorem 4.1 implies that for all $(x, y)\in O$,

$$F_3(x,y,\nu(x,y)) = \frac{\partial\nu}{\partial x}(x,y)F_1(x,y,\nu(x,y)) + \frac{\partial\nu}{\partial y}(x,y)F_2(x,y,\nu(x,y)),$$

F_1, F_2 (= 0), F_3, being the components of F. Hence:

$$-v(x,y)+x^2+\frac{\partial v}{\partial x}(x,y)(yx+x^3)=0.$$

By successive derivations w.r.t. x,

$$0=-\frac{\partial v}{\partial x}(x,y)+2x+\frac{\partial^2 v}{\partial x^2}(x,y)(yx+x^3)+\frac{\partial v}{\partial x}(x,y)(y+3x^2),$$

$$0=-\frac{\partial^2 v}{\partial x^2}(x,y)+2+\frac{\partial^3 v}{\partial x^3}(x,y)(yx+x^3)+2\frac{\partial^2 v}{\partial x^2}(x,y)(y+3x^2)$$
$$+6\frac{\partial v}{\partial x}(x,y)x,$$

and for $n \geq 3$,

$$0=-\frac{\partial^n v}{\partial x^n}(x,y)+\frac{\partial^{n+1} v}{\partial x^{n+1}}(x,y)(yx+x^3)+n\frac{\partial^n v}{\partial x^n}(x,y)(y+3x^2)$$
$$+3n(n-1)x\frac{\partial^{n-1} v}{\partial x^{n-1}}(x,y)+n(n-1)(n-2)\frac{\partial^{n-2} v}{\partial x^{n-2}}(x,y).$$

Hence, for all $y \in]-r,r[$, we get

$$0=(y-1)\frac{\partial v}{\partial x}(0,y),\ 2=(1-2y)\frac{\partial^2 v}{\partial x^2}(0,y),$$

and for $n \geq 3$,

$$\frac{\partial^{n-2} v}{\partial x^{n-2}}(0,y)=\frac{(1-ny)}{n(n-1)(n-2)}\frac{\partial^n v}{\partial x^n}(0,y).$$

From this, we get that $\frac{\partial^{2p+1} v}{\partial x^{2p+1}}(0,y)=0$ for all $y \in]-r,r[$, all $p \in \mathcal{N}$, and that for all $y \in]-r,r[$, all $q \in \mathcal{N}$, $q \geq 1$,

$$2=\frac{\prod_{k=1}^{q}(1-2ky)}{(2q)!(q-1)!2^{q-1}}\frac{\partial^{2q} v}{\partial x^{2q}}(0,y).$$

As soon as $q > \frac{1}{2r}$, evaluating at $y=\frac{1}{2q}$, we get a contradiction. Hence, v can be at most of class C^ρ, where ρ is the integer part of $\frac{1}{2r}$.

Remark 4.2. Obviously, local center manifolds of F at x_0 are not unique. However, one could think that the germ of a local center manifold of F at x_0 is unique. This is not true, as the following example shows.

Example 4.2. Let $M = R^2$, with canonical coordinates (x, y), $x_0 = (0, 0)$, $F = -x^3 \frac{\partial}{\partial x} - y \frac{\partial}{\partial y}$. For any $a, b \in R$, set:

$$\nu_{a,b} : R \to R,$$

$$\nu_{a,b}(x) = \begin{cases} ae^{-\frac{1}{2x^2}} \text{ if } x > 0, \\ 0 \text{ if } x = 0, \\ be^{-\frac{1}{2x^2}} \text{ if } x < 0. \end{cases}$$

For any a, b, the graph of $\nu_{a,b}$ is a local center manifold of class C^∞ of F at $x_0 = (0, 0)$.

4.2. Applications of the Center Manifold Theorem

Let M, F, x_0, LF, n, k be as in Section 4.1. Let (U, N) be a C^r local center manifold of F at x_0.

We can then define a new vector field F_N, the **reduced vector field**: F_N is the restriction of F to N. By Property (ii) of Theorem 4.1, it is a vector field on the manifold N, of class C^{r-1}. It has x_0 as a stationary point. Essentially, the local properties of F at x_0 are entirely determined by those of F_N at x_0:

Theorem 4.2.

(i) If the restriction of LF to h has an eigenvalue with positive real part, x_0 is an unstable point of F.

(ii) If the restriction of LF to h has no eigenvalue with positive real part, then x_0 is stable (asymptotically stable) for F if and only if x_0 is stable (respectively asymptotically stable) for F_N in N.

We do not need point (i) in the book. The proof of (ii) is given in [9].

Solutions to Part I Exercises

Some of the exercises in the theoretical part (Part I) of this book are important for the development of our observability theory. Hence, we correct completely most of them. For a few exercises, we only give a reference: The results are rather clear, but they involve long developments, similar to those in the book. We leave to the reader the exercises of the second part of the book, which contains only applications of our observability theory.

1. Chapter 2

1.1. Exercise 5.1

(a) Δ_Σ is f_u–invariant for all $u \in U$ (i.e. $[\Delta_\Sigma, f_u] \subset \Delta_\Sigma$): $[\Delta_\Sigma, f_u] \subset \Delta_\Sigma$ iff $L_{[X,\ f_u]}\varphi = 0$ for all $\varphi \in \Theta^\Sigma$ and all vector fields X that are sections of Δ_Σ. Otherwise $L_{[X,\ f_u]}\varphi = L_X L_{f_u}\varphi - L_{f_u} L_X \varphi = 0$, because $L_X\varphi = 0$ by definition of Δ_Σ, and $L_X L_{f_u}\varphi = 0$ because Θ^Σ is stable by L_{f_u}.

(b) The question is local. Because Δ_Σ is regular and nontrivial, we can find coordinates $x^1, \ldots, x^p, x^{p+1}, \ldots, x^n$, $p \neq 0$, such that $\Delta_\Sigma = span\{\frac{\partial}{\partial x^1}, \ldots, \frac{\partial}{\partial x^p}\}$. The definition of Δ_Σ implies that

$$\frac{\partial h}{\partial xi} = 0,\ 1 \leq i \leq p.$$

Point (a) above implies that

$$\frac{\partial f_{u,j}}{\partial x^i} = 0,\ 1 \leq i \leq p,\ \ p+1 \leq j \leq n.$$

Hence, we have the following (local) normal form for Σ:

$$\begin{aligned}
\dot{x}^1 &= f_1(x, u),\\
&\cdot\\
&\cdot\\
\dot{x}^p &= f_p(x, u),\\
\dot{x}^{p+1} &= f_{p+1}(x^{p+1}, \ldots, x^n, u),\\
&\cdot\\
&\cdot\\
\dot{x}^n &= f_n(x^{p+1}, \ldots, x^n, u), \quad y = h(x^{p+1}, \ldots, x^n, u).
\end{aligned} \tag{192}$$

The result follows immediately, because the flow $\varphi_t(x)$ corresponding to (192) for any $u(.)$, maps two points x_1 and x_2 in some leaf $\{x^{p+1} = c_{p+1}, \ldots, x^n = c_n\}$ into two points in the same leaf of Δ_Σ.

1.2. *Exercise 5.2*

(a) $\Theta^\Sigma = span\{(L_f)^k h\ ; 0 \leq k\}$. Let $x_1, x_2 \in X$, $x_1 \neq x_2$. By assumption, there is a k such that $(L_f)^k h(x_1) \neq (L_f)^k h(x_2)$. Hence, if $y_1(t)$ and $y_2(t)$ denote the output functions associated with the initial conditions x_1 and x_2 respectively, then,

$$\frac{\partial^k y_1}{\partial t^k}(0) \neq \frac{\partial^k y_2}{\partial t^k}(0).$$

This shows that the "*initial – state→ output – trajectory*" mapping is injective.

(b) In the analytic situation, the output function $y(t)$ is analytic function of the time, and we have the Taylor expansion, valid for small t:

$$y(t) = \sum_{k=0}^{\infty} (L_f)^k h(x) \frac{t^k}{k!}.$$

If the system is observable, two (analytic) output functions $y_1(t)$ and $y_2(t)$ cannot have the same Taylor expansion at $t = 0$, otherwise, they would coincide on the interval $[0, \min(e(x_1), e(x_2))[$. Hence, Θ^Σ separates the points.

1.3. *Exercise 5.3.* $X = R$, $d_u = 1$, $d_y = 1$, $U = [-1, 1]$,

$$(\Sigma)\begin{cases} \dot{x} = 0,\\ y = x\,\varphi(u), \end{cases}$$

where φ is C^∞, flat at $u = 0$, but nonidentically zero.

2. Chapter 3

2.1. *Exercise 4.1.* Because Σ has a uniform canonical flag, by Theorem 2.1, Σ can be put everywhere locally under the normal form (20). A direct computation using this normal form shows that the map $S\Phi_n^\Sigma$ is an injective immersion. The considerations at the end of Section 3 of Chapter 2 give the result.

2.2. *Exercise 4.2.*

(a) If two output trajectories $y_1(t)$, $y_2(t)$ of the system coincide on some interval $[0, \varepsilon]$, then the functions $\sqrt[3]{y_1}(t)$, $\sqrt[3]{y_2}(t)$ coincide also. However, these functions are output functions of the (observable) linear system:

$$\dot{x}^1 = x^2, \ \dot{x}^2 = 0, \ y = x^1.$$

This shows that the system is observable.

(b) The first variation has the following expression:

$$\begin{aligned} \dot{x}^1 &= x^2, \ \dot{x}^2 = 0, \\ \dot{\xi}^1 &= \xi^2, \dot{\xi}^2 = 0, \\ \eta &= 3\,(x^1)^2\,\xi^1. \end{aligned}$$

If we start from $x(0) = 0$, for all initial conditions $\xi(0)$, the output $\eta(t)$ of the first variation is identically zero. This shows the result.

2.3. *Exercise 4.3.*

1. Because Σ has a uniform canonical flag by Theorem 2.1, it can be put everywhere locally under the normal form (20). In this special case, this gives only the condition:

$$dL_{f_u}h \wedge dh \neq 0.$$

Therefore, Σ has a uniform canonical flag on a neighborhood of $x \in X$ iff $dL_{f_u}h(x) \wedge dh(x) \neq 0$. The point x^0 being fixed, the mapping $(f, h, u) \to dL_{f_u}h(x^0) \wedge dh(x^0)$ is continuous for the C^2 topology on S. The result follows.

2. If $n > 2$, let us assume that $d_u = 1$, and $U = [-1, 1]$. The distribution $D_1(u)$ is given by

$$D_1(u) = \ker dh \cap \ker dL_{f_u}h,$$

and it has to be independent of u. Moreover, by the normal form, $dL_{f_0}h \wedge dh \neq 0$ and $dL_{f_u}h \wedge dL_{f_0}h \wedge dh = 0$: if not, the codistribution annihilating $D_1(u)$ depends on u, which is impossible. Differentiating w.r.t u at $u = 0$, shows that, for all $k \geq 1$ and for all $u \in U$,

$$dL_{(\frac{\partial}{\partial u})^k f_u} h_{|u=0} \wedge dL_{f_0}h \wedge dh = 0. \tag{193}$$

Let us work in the open subset $\mathcal{O}$ of the set S^r of C^r systems, r large, such that $dL_{f_0}h \wedge dh \neq 0$ on Ω, Ω a fixed compact subset of X.

A direct simple computation shows that the subset of $J^{K+1}S$ defined by the relation (193) for all $k \leq K$ is a closed submanifold of codimension $(n-2)K$. Because $n > 2$, this codimension can be made arbitrarily large, increasing K. By the transversality theorems, the set of C^r systems that miss this condition at all $x \in \Omega$ is open dense. Taking $\Omega = \{x_0\}$ shows the result.

2.4. Exercise 4.4. In this special case, the conditions for the uniformity of the canonical flag are those of Theorem 4.1. The conditions of this theorem, in the two-dimensional case, can be expressed as follows:

$$(a)\ dh \wedge dL_f h \neq 0,$$

$$(b)\ dL_g h \wedge dh = 0.$$

Again, let us work on the open set $\mathcal{O}$ of systems such that (a) is true at all $x \in \Omega$, Ω a fixed compact subset of X.

Then, for $\Sigma \in \mathcal{O}$, we have a vector field Z on an open neighborhood of Ω, uniquely defined by (i) $L_Z h = 0$, (ii) $L_Z L_f h = 1$. The condition (b) implies that $L_Z^k(dL_g h \wedge dh) = 0$ on Ω, for all k. A simple computation shows that this condition for $k \leq K$ defines a closed submanifold of $J^{K+2}S$ of codimension K. Hence, the set of systems that miss this condition at all $x \in \Omega$ is open, dense, for $K > 2$. Taking $\Omega = \{x_0\}$ gives the result.

2.5. Exercise 4.5.

$$\frac{\partial y}{\partial x} = \sin(u)^2 + \sin(x(1+u^2)^{\frac{1}{2}})^2.$$

Therefore, $\frac{\partial y}{\partial x} = 0$ iff $u = k\pi$ and $x = \frac{m\pi}{(1+k^2\pi^2)^{\frac{1}{2}}}$, for all integers k, m.

This last expression defines a countable dense set M on $X = R$. However, this set is not invariant by the dynamics of the system. For any input $\hat{u}$, this shows that any output trajectory of the first variation of the system, starting from a $\xi(0) \notin s_0(X)$, where s_0 is the zero section of $\pi : TX \to X$, is nonzero as an element of $L^\infty[0, T]$, for $T > 0$. The system is uniformly infinitesimally observable. However, condition 1 of Theorem 3.2 is violated: M is dense.

2.6. Exercise 4.6. Let us denote by $M_k, k = 0, \dots, n-1$, the subset of $X = R^n$ on which $dh \wedge dL_f h \wedge \dots \wedge dl_f^k h(x) \neq 0$. M_k is open. The C^∞ version of Theorem 4.1 is as follows.

Theorem 2.1. *1. If Σ_A is **observable**, then M_{n-1} is open, dense, and, on each open subset $Y \subset X \backslash M_{n-1}$, such that the restriction $\Phi_{|Y}$ is a diffeomorphism, $\Phi_{|Y}$ maps $\Sigma_{A|Y}$ into a system $\bar{\Sigma}_A$ of the form:*

$$\bar{\Sigma}_A, \ \dot{x} = \begin{pmatrix} \dot{x}_1 \\ \dot{x}_2 \\ \cdot \\ \cdot \\ \cdot \\ \dot{x}_{n-1} \\ \dot{x}_n \end{pmatrix} = \begin{pmatrix} x_2 \\ x_3 \\ \cdot \\ \cdot \\ \cdot \\ x_n \\ \varphi(x) \end{pmatrix} + u \begin{pmatrix} g_1(x_1) \\ g_2(x_1, x_2) \\ \cdot \\ \cdot \\ \cdot \\ g_{n-1}(x_1, \dots, x_{n-1}) \\ g_n(x) \end{pmatrix}. \quad (194)$$

$$y = x_1.$$

2. Conversely, if Ω is an open subset of R^n on which the system Σ has the form $\bar{\Sigma}_A$, then the restriction $\Sigma_{|\Omega}$ is observable.

Proof. Once we have shown that M_{n-1} open, dense (in place of being analytic, with codimension (1), the remainder of the proof is completely similar. Let us show by induction on n that M_{n-1} is dense.

First, if $dh = 0$ on an open connected subset, then h is constant on this subset, and Σ_A is not observable. Hence, M_0 is dense. Assume that M_k, $k < n-1$, is open, dense. Let $U \subset X$, U open, be such that $dh \wedge dL_f h \wedge \dots \wedge dl_f^{k+1} h(x) = 0$ on U. Then, on a $V \subset U$, V open, we can chose coordinates x_i, $x_1 = h, \dots, x_{k+1} = L_f^k h, x_{k+2}, \dots, x_n$, with $L_f^{k+1} h(x) = \varphi(x_1, \dots, x_{k+1})$, where φ is smooth. With the same reasoning as in the proof of Theorem 4.1, $L_f^m h(x)$ depends only on $x_1, \dots, x_{k+1}$ on V for all $m > 0$. Therefore, considering the "drift system" (uncontrolled):

$$(\Sigma_d) \begin{cases} \dot{x} = f(x), \\ y = h(x), \end{cases}$$

we can apply the result of Exercise 5.1 of Chapter 2, to Σ_d restricted to V, because on V, the foliation Δ_{Σ_d} is regular and nontrivial. The system Σ_d

can be put in the normal form (192), and hence is not observable, which is contradiction. This ends the proof. ■

2.7. Exercise 4.7. Let us apply Theorem 4.1 to the special case of the bilinear system B. If B is observable, it implies that the linear forms $Cx, CAx, \ldots,$ $CA^{n-1}x$ form a linear coordinate system on R^n. In this coordinate system, B is still bilinear, and has the **bilinear** normal form (194). This shows the result.

2.8. Exercise 4.8.

1. Given two systems Σ, Σ':

$$(\Sigma)\begin{cases} \dot{x} = f(x, u), \\ y = h(x, u), \end{cases}, (\Sigma')\begin{cases} \dot{z} = f'(z, u), \\ y = h'(z, u), \end{cases},$$

on two manifolds X, Z, with control spaces U and output spaces R^{d_y}, the most natural notion of an immersion of Σ into Σ' is:

Definition 2.1. $\tau : X \to Z$ is called an immersion of Σ into Σ' if:

1) τ is a smooth mapping (we do not require that τ is immersive, but nevertheless, the traditional terminology is "immersion"),
2) for all $u \in L^\infty(U)$, all $x \in X$, $e_\Sigma(u, x) \leq e_{\Sigma'}(u, \tau(x))$, and, for all $t \in$ $[0, e_\Sigma(u, x)[$:

$$P_\Sigma(u, x)(t) = P_{\Sigma'}(u, \tau(x))(t).$$

Here, $e_\Sigma, e_{\Sigma'}, P_\Sigma, P_{\Sigma'}$ denote the explosion-time functions and the input-output mappings of Σ, Σ'.

2. Assume that a system Σ can be immersed into a "state affine" one Σ':

$$(\Sigma)\begin{cases} \dot{x} = f(x, u), \\ y = h(x, u), \end{cases}, (\Sigma')\begin{cases} \dot{z} = A(u)z + b(u) = B(u, z), \\ y = C(u)z + d(u). \end{cases}$$

Considering piecewise constant control functions, for all $u_1, \ldots,$ $u_{r+1} \in U$, all $x \in X$,

$$L_{f_{u_1}\ldots f_{u_r}} h_{u_{r+1}}(x) = [L_{B_{u_1}\ldots B_{u_r}}(C(u_{r+1})z + d(u_{r+1}))] \circ \tau(x).$$

The right-hand terms in the previous expressions generate a finite dimensional vector space. Hence, the same applies to the left-hand terms. The space Θ^Σ is finite dimensional.

3. Conversely, assume that $\dim(\Theta^{\Sigma}) = m < \infty$. Then, considering the following "state-linear" system Σ' on $(\Theta^{\Sigma})^*$, dual space of Θ^{Σ}:

$$(\Sigma')\begin{cases} \dot{\varphi} = A(u)(\varphi), \\ y = C(u)(\varphi), \end{cases}$$

where
- $A(u):(\Theta^{\Sigma})^* \to (\Theta^{\Sigma})^*$, $A(u)(\varphi)(\psi) = \varphi(L_{f_u}(\psi))$, for $\psi \in \Theta^{\Sigma}$ ($A(u) = L'_{f_u}$, transpose of "L_{fu} restricted to Θ^{Σ}"),
- $C(u)(\varphi) = \varphi(h_u)$,
- $\tau : X \to (\Theta^{\Sigma})^*$, $\tau(x)(\psi) = ev_x(\psi) = \psi(x)$, for $\psi \in \Theta^{\Sigma}$.

Then, the mapping τ is an immersion of Σ into Σ'.

4. If moreover Σ is control affine, the system Σ' above is control affine, hence bilinear.

3. Chapter 4

3.1. Exercise 0.1. The basic example comes from Hirsch [26]. It concerns mappings from Z to R^n. Consider $f : Z \to R^n$, the image of which is the set of points with rational coordinates. Let $g : Z \to R^n$, any map such that

$$\| f(i) - g(i) \| < \frac{1}{|i|},$$

for $i \neq 0$. The image of g is dense, hence, g is not an embedding.

3.2. Exercise 2.1. Let us denote here the circle by T, and the set of C^r systems on T, with control space U and output space R, by S^r. The cyclic coordinate on T is θ.

(i) Let $\Sigma^0 = (h^0(\theta, u), f^0(\theta, u))$ be given, $\Sigma^0 \in S^{\infty}$, with

$$\frac{\partial h^0}{\partial \theta}(0,0) = 0, \quad \frac{\partial^2 h^0}{\partial \theta^2}(0,0) \neq 0, \quad \frac{\partial^2 h^0}{\partial \theta \partial u}(0,0) \neq 0. \tag{195}$$

There exists $a > 0$, and neighborhoods $V_{\Sigma^0} \subset S^2$, $W_0 \subset T$, $\theta_0 : V_{\Sigma^0} \to W_0$, C^2-continuous, such that

(i)-1: $\frac{\partial h}{\partial \theta}(\theta_0(\Sigma), 0) = 0$, for all $\Sigma \in V_{\Sigma^0}$,

(i)-2: $|\frac{\partial^2 h}{\partial \theta \partial u}(\theta, 0)| > a$, for all $\theta \in W_0$, for all $\Sigma \in V_{\Sigma^0}$.

The statement (i)-1 is just a consequence of the implicit function theorem.

(ii) There exists $\tilde{U}_{N-1} : (\theta, \Sigma) \in W_0 \times (V_{\Sigma^0} \cap S^N) \to \tilde{U}_{N-1}(\theta, \Sigma) \in R^{(N-1)d_y}$, C^N-continuous, such that $d_\theta \pi_2 \Phi_N^{\Sigma}(\theta, 0, \tilde{U}_{N-1}(\theta, \Sigma)) = 0$. Here,

$\pi_2 : R^{Nd_y} \to R^{(N-1)d_y}$ is the projections that forgets $y : \pi_2(y, \tilde{y}_{N-1}) = \tilde{y}_{N-1}$.

(ii) results from (i)-2 and from the following trivial computation:

$$d_\theta \dot{y} = \frac{\partial}{\partial\theta}\left(L_{f_u} h_u + \frac{\partial h}{\partial u}\dot{u}\right),$$

$$\cdot$$

$$\cdot$$

$$d_\theta y^{(i)} = \frac{\partial}{\partial\theta}\left(\Psi_i(\Sigma, \tilde{u}_{i-1}) + \frac{\partial h}{\partial u} u^{(i)}\right),$$

where Ψ_i is a universal function depending on $(\frac{\partial}{\partial u})^{j_1} L_{f_u}^{k_1} \ldots .(\frac{\partial}{\partial u})^{j_r} L_{f_u}^{k_r} (\frac{\partial}{\partial u})^{j_{r+1}}$ $h_u(x)$, $\tilde{u}_{i-1}$, with $\Sigma j_s + \Sigma k_s \leq i$.

Any system Σ in $V_{\Sigma^0} \cap S^N$ has the properties $(i) - 1$, $(i) - 2$ above. With (ii) above, we can compute $\tilde{u}_{N-1}$ so that

$$d_\theta \Phi_N^\Sigma(\theta_0(\Sigma), 0, \tilde{U}_{N-1}(\theta_0(\Sigma), \Sigma)) = 0,$$

which means that Φ_N^Σ is not immersive at $(\theta_0(\Sigma), 0, \tilde{U}_{N-1}(\theta_0(\Sigma), \Sigma))$. By taking for Σ^0, Σ^1 in the statement of the exercise, we get the result.

3.3. Exercise 2.2.

(a) $\tilde{U}_{N-1}(0, \Sigma_\varepsilon^1)$ is uniquely defined for all $\varepsilon > 0$, as above (i.e., taking the unique solution of $d_\theta \pi_2 \Phi_N^{\Sigma_\varepsilon^1}(0, 0, \tilde{U}_{N-1}(0, \Sigma_\varepsilon^1)) = 0$). Direct computations show that we can choose ε small enough for $\tilde{U}_{N-1}(0, \Sigma_\varepsilon^1) \in (I_{\frac{B}{2}})^{(N-1)d_u}$.

(b) Let us take Σ_ε^1 for Σ^0 in (i) and (ii) of the proof of the previous exercise. If Σ is C^N-close to Σ_ε^1, then $\theta_0(\Sigma)$ is close to zero, and $\tilde{U}_{N-1}(\theta_0, \Sigma)$ is close to $\tilde{U}_{N-1}(0, \Sigma_\varepsilon^1)$ by (ii) in the previous exercise.

Therefore, for Σ sufficiently close (C^N) to Σ_ε^1, $\tilde{U}_{N-1}(\theta_0, \Sigma) \in (I_B)^{(N-1)d_u}$. This shows that Φ_N^Σ is not immersive at $(\theta_0, 0, \tilde{U}_{N-1}(\theta_0, \Sigma))$, and $\tilde{U}_{N-1}(\theta_0, \Sigma) \in (I_B)^{(N-1)d_u}$. This is what we wanted to show.

3.4. Exercise 2.3.

(a) $S\Phi_2^{\Sigma_0^2}(\theta, u, \dot{u}) = (\cos(\theta), 0, -\sin(\theta), 0, u, \dot{u})$. $S\Phi_2^{\Sigma_0^2}$ is immersive.

(b) for ε nonzero, we are in the context of Exercise 2.1. Hence, for ε small enough, $S\Phi_k^{\Sigma_\varepsilon^2}$ is not an immersion.

3.5. *Exercise 2.4.* Assume that $\Sigma \to S\Phi_N^\Sigma$ is continuous. Because the immersions are open in

$$\left(C^{r-N+1}\left(X \times U \times R^{(N-1)d_u}, R^{Nd_y} \times U \times R^{(N-1)d_u}\right), \text{ Whitney topology}\right),$$

for $r - N + 1 > 1$, the set of Σ such that $S\Phi_N^\Sigma$ is an immersion is open. This is false, by the previous exercise.

3.6. *Exercise 4.1.* This example is a personal communication of H. J. Sussmann. Let $X = R^2$, $U = [-1, 1]$, $d_y = 3$.

$$(\Sigma) \begin{cases} \dot{x}_1 = 1, \\ \dot{x}_2 = u, \\ y_1 = \varphi(x_1, x_2)x_1, \;\; y_2 = \varphi(x_1, x_2)x_2, \;\; y_3 = \varphi(x_1, x_2), \end{cases}$$

with

$u_0 : [0, 1] \to U$, a function which is everywhere C^k, and nowhere C^{k+1},

$v(t) = \int_0^t u_0(\tau)d\tau$,

$\gamma_1 : [0, 1] \to X$, $\gamma_1(t) = (t, v(t))$,

$\gamma_2 : [0, 1] \to X$, $\gamma_2(t) = (t, v(t) + 1)$,

$\varphi : X \to R$ is a smooth function, nonzero everywhere except on $\gamma_1([0, 1]) \cup \gamma_2([0, 1])$ (such a function does exist because this last set is closed).

3.7. *Exercise 6.1.* This is a special case of Theorem 2.3.

3.8. *Exercise 6.2.* See [3].

3.9. *Exercise 6.3.* See [31].

4. Chapter 5

4.1. *Exercise 3.1.* The observation space Θ^Σ of such a system Σ is finite dimensional: It is contained in the vector space of polynomials of a fixed degree. By Exercise 4.8 in Chapter 3, it can be immersed in a linear (uncontrolled) system. This system can be taken to be observable. Any observable linear system satisfies the $ACP(N)$ for some N. Nevertheless, the multiplicity may be infinite, as the following exercise shows.

4.2. Exercise 3.2.

1. $y = h(x) = x_1(x_1^2 + x_2^2)$, $\dot{y} = L_{Ax}h = x_2(x_1^2 + x_2^2)$. These two functions separate the points on R^2.
2. In fact, for $N \geq 2$, $\mathfrak{R}_N = (\Phi_2^\Sigma)^*(O_{y_0})$, and $(\Phi_2^\Sigma)^*(\mathrm{m}(O_{y_0})).O_{x_0} \subset (x_1^2 + x_2^2).O_{x_0}$, which has infinite codimension in O_{x_0}.

4.3. Exercise 4.1.

1. $y = x_1$, $\dot{y} = x_2^3 - x_1$. Any two distinct initial conditions are distinguished immediately.

2a. $y = h = x_1$, $\dot{y} = L_{f_u}h = x_2^3 - x_1$, this shows that $\check{\mathfrak{R}}_2 = \mathfrak{R}_2 = \{G(u, x_1, x_2^3)\}$.

2b. $\ddot{y} = x_1 - x_2^3 + 3\,u\,x_2^6 + 3\,x_2^{10}$, $L_{f_u}(x_2^3) = 3\,x_2^{10} + 3\,x_2^6\,u$, this shows that:

$$\check{\mathfrak{R}}_3 = \mathfrak{R}_3 = \left\{G\left(u, x_1, x_2^3, x_2^{10}\right)\right\}.$$

2c.

$$y^{(3)} = -x_1 + x_2^3 - 3ux_2^6 + 3u'x_2^6 + 18u^2x_2^9 - 3x_2^{10} + 48ux_2^{13} + 30x_2^{17},$$

$$L_{f_u}\left(x_2^{10}\right) = 10\left(x_2^{17} + u\,x_2^{13}\right),$$

hence,

$$\check{\mathfrak{R}}_4 = \mathfrak{R}_4 = \left\{G\left(u, x_1, x_2^3, x_2^{10}, x_2^{17}\right)\right\}.$$

2d.

$$L_{f_u}\left(x_2^{17}\right) = 17\left(x_2^{24} + u\,x_2^{20}\right),$$

this shows that

$$\check{\mathfrak{R}}_5 = \check{\mathfrak{R}}_4 = \mathfrak{R}_4 = \check{\mathfrak{R}}.$$

Hence, Σ satisfies the $ACP(4)$, by Theorem 4.1.

3. $P = 3u''y^2 - 12u\,u'y^3 + 510u^3\,y^4 + 510y^8 + 22u'y\,y' + 6u''y\,y' - 494u^2\,y^2\,y' - 36u\,u'\,y^2\,y' + 2040u^3\,y^3y' + 4080y^7\,y' + 126u\,y'^2 + 22u'\,y'^2 + 3u''\,y'^2 - 988u^2y\,y'^2 - 36u\,u'\,y\,y'^2 + 3060u^3\,y^2\,y'^2 + 14280y^6\,y'^2 - 494u^2\,y'^3 - 12u\,u'\,y'^3 + 2040u^3\,y\,y'^3 + 28560y^5\,y'^3 + 510u^3\,y'^4 + 35700y^4\,y'^4 + 28560y^3\,y'^5 + 14280y^2\,y'^6 + 4080y\,y'^7 + 510y'^8 + 22u'\,y\,y'' - 494u^2\,y^2\,y'' + 252u\,y'\,y'' + 22u'\,y'\,y'' - 988u^2\,y\,y'\,y'' - 494u^2\,y'^2\,y'' + 126u\,y''^2 - y'''$.

4.4. Exercise 5.1.

1. If two trajectories $x(t)$, $\tilde{x}(t)$ produce the same outputs $y(t)$, $\tilde{y}(t)$, then $y(0) = \tilde{y}(0)$, $y'(0) = \tilde{y}'(0)$, $y''(0) = \tilde{y}''(0)$, $y'''(0) = \tilde{y}'''(0)$. This implies:

$$\cos(\alpha\, x_0) = \cos(\alpha\, \tilde{x}_0), \ \cos(x_0) = \cos(\tilde{x}_0),$$
$$\sin(\alpha\, x_0) = \sin(\alpha\, \tilde{x}_0), \ \sin(x_0) = \sin(\tilde{x}_0).$$

Because α is irrational, this implies that $x_0 = \tilde{x}_0$. Σ is observable.
2. The observation space is finite dimensional: It is generated by $\cos(x)$, $\cos(\alpha x)$, $\sin(x)$, $\sin(\alpha x)$. The $ACP(4)$ holds.
   Let us compare with Exercise 3.1: In Exercise 3.1, h, the output function, is a polynomial mapping. In this exercise, h is an almost periodic function on R. This is a special case of a general situation, for certain left invariant systems on Lie Groups.
3. The image of R by Φ_M^Σ is a relatively compact set Ω_M, for all M. Assume that $x = \varphi(y, \bar{y}_{M-1})$ for some M, for a continuous mapping φ defined on R^M. Then, $\varphi(\Omega_M) = R$, which is impossible because φ is globally defined and Ω_M is relatively compact.

4.5. Exercise 5.2.

1. The observation space of Σ contains x_1, x_2^3. By Exercise 5.2 in Chapter 2, Σ is differentially observable: the mapping Φ_2^Σ is injective.
2. If $f(x_1, x_2) = 1$, Φ_4^Σ is an embedding. If $f(x_1, x_2) = x_2$, for $N \geq 2$, $\Re_N = \{G(x_1, x_2^3)\}$. It has not full rank at $x = 0$. Then, Φ_M^Σ is not an immersion for any M.

4.6. Exercise 5.3.

1. $S\Phi_2^\Sigma(x_1, x_2, u, u') = (x_1, x_2^3 - x_1, u, u')$. It is an injective mapping. Σ is differentially observable of order 2.
2. An easy induction shows that $\frac{\partial y^{(N)}}{\partial x_2}\Big|_{x=0} = 0$ for all N. Then, Σ is not strongly differentially observable of any order.

4.7. Exercise 5.4. Σ satisfies the $PH(n)$ at each point by Exercise 4.1 in Chapter 3. Hence, it satisfies the $PH(N)$ or equivalently the $ACP(N)$ at each point, for $N \geq n$. Theorem 5.2 shows the result.

4.8. Exercise 6.1. The point $x = 0$ is a fixed point for the system, and the axis $x_2 = 0$ is invariant. Writing the dynamics in a control affine way, $\dot{x} = f(x) + ug(x)$, and denoting the Lie bracket of f and g by $e(x) = [f, g](x)$, we see that g and e are independent, except if $x_2 = 0$. If $x_2 = 0$ but $x_1 \neq 0$, $f(x)$ is nonzero. Hence, the system is not controllable in the weak sense: $X = R^2$ is partitioned into five orbits.

5. Chapter 6

5.1. Exercise 1.1. Let us assume that $\lim_{t\to+\infty} \#E(t, z_0) \geq 2$. Then, by compactness of $Cl(\Omega)$, $Cl(\Omega')$, U, I_B, there exists $t_n \to +\infty$, $\hat{x}_n^1, \hat{x}_n^2 \in E(t_n, z_0)$, $\hat{x}_n^1 \neq \hat{x}_n^2$, $x_n = x(t_n) \to x^*$, $\hat{x}_n^1 \to x^*$, $\hat{x}_n^2 \to x^*$, $\underline{u}_N^n = (u_n^{(0)}, \tilde{u}_{N-1}^n) \to \underline{u}_N^*$, $\eta_n = \eta(t_n) \to \eta^* = \Phi_N^\Sigma(x^*, \underline{u}_N^*)$.

Also, d_X denoting again the differential with respect to the x variable only,

$$d_X \left\| \Phi_N^\Sigma(\hat{x}_n^i, \underline{u}_N^n) - \eta_n \right\|^2 = 0, \text{ for } i = 1, 2,$$

because, in coordinates, for all V,

$$\left\| \Phi_N^\Sigma(\hat{x}_n^i, \underline{u}_N^n) - \eta_n \right\|^2 \leq \left\| \Phi_N^\Sigma(\hat{x}_n^i + \lambda V, \underline{u}_N^n) - \eta_n \right\|^2,$$

if n is sufficiently large, and $\lambda > 0$ is small enough.

But, by the implicit function theorem, and by the fact that $S\Phi_N^\Sigma$ is immersive, the equation $d_X \| \Phi_N^\Sigma(\hat{x}, \underline{u}_N^n) - \eta_n \|^2 = 0$ has a single solution $\hat{x}_n$ in a neighborhood of $(x^*, \underline{u}_N^*, \eta^*)$.

5.2. Exercise 1.2. First, for a fixed $\gamma = (u^{(0)}, \tilde{u}_{N-1})$, $\|\Phi_N^\Sigma(., \gamma) - \Phi_N^\Sigma(., \gamma)\|$ defines a distance $d_\gamma(.,.)$, because $S\Phi_N^\Sigma$ is injective. For x and y fixed, the distance $d_\gamma(x, y) : G \to R_+$ is continuous, where G is the set of values of γ. The set G is compact. Hence, $d(x, y) = \sup_\gamma d_\gamma(x, y) < +\infty$. The other properties of a distance are easy to check.

5.3. Exercise 1.3. For $N \geq 4$, Φ_N^Σ is an injective immersion, and the image Ω_N of $X = R$ by Φ_N^Σ is contained in a 4-dimensional subspace V_N of R^N. Moreover, the closure of Ω_N is a 2-dimensional torus T_N in V_N, and Ω_N is dense in T_N.

Let x^* be any point of $X = R$, and $y^* = \Phi_N^\Sigma(x^*)$. An open neighborhood U_{y^*} of y^* in R^N has a trace V_{y^*} on V_N, and Ω_{y^*} on Ω_N. The set Ω_{y^*} contains a sequence $\Phi_N^\Sigma(x_n)$, $x_n \to +\infty$ in $X = R^N$. Hence, an open interval $]x^* - a, x^* + a[$, $a > 0$, is not an open neighborhood of x^* in the topology defined by the observability distance.

5.4. *Exercise 1.4.* Because X is compact and Φ_N^Σ is an injective smooth mapping, Φ_N^Σ is an homeomorphism onto its image, which is closed (Lemma 5.1 of Chapter 5). This shows the result.

5.5. *Exercise 1.5.* Let d, δ be two Riemannian metrics over X. It is a standard fact that smooth mappings between Riemannian manifolds are locally Lipschitz. As a consequence, because C is compact, there is an open neighborhood U^C of ΔC (the diagonal of $C \times C$) in $X \times X$, and a real $\alpha > 0$, such that:

$$d(x, y) \leq \alpha\ \delta(x, y),\ (x, y) \in U^C.$$

Let V^C be the complement of U^C in $X \times X$, and $W^C = V^C \cap (C \times C)$. Then, V^C is closed and W^C is compact.

Hence, $\delta : W^C \to R_+$, reaches its minimum $\delta_m > 0$. Let d_M be the (finite) diameter of C for the metric d.

For $(x, y) \in W^C$, $d(x, y) \leq d_M = \frac{d_M}{\delta_m}\delta_m \leq \frac{d_M}{\delta_m}\delta(x, y)$. Therefore, for $(x, y) \in C \times C$, $d(x, y) \leq \sup(\alpha, \frac{d_M}{\delta_m})\delta(x, y)$. This shows the result.

5.6. *Exercise 1.6.* First, on any Riemannian manifold, there is a complete Riemannian metric (see [34], p. 12, or use the Whitney's embedding theorem to construct the metric). Let us fix the manifold X, and choose such a complete Riemannian metric g on X, with associated distance d.

Fix $x_0 \in X$, and set $f(x) = d(x_0, x)$. Let $\varphi : X \to R_+$ be a C^∞ function, $\varphi(x) > \sup(1, f(x))$. Such a function can be easily constructed, using a partition of unity. Let g' be the Riemannian metric $g' = e^{-2\varphi}g$. The associated distance function is δ.

The diameter of X for the distance δ is finite: Let $x \in X$, and $\gamma : [0, t] \to X$ be a geodesic from x_0 to x, relative to g, parametrized by the arclength, so that $f(x) = d(x_0, x) = t$. Such a geodesic does exist ([34], p. 126).

Then, denote by $L_\delta(\gamma)$ the length of the curve γ with respect to the metric g':

$$L_\delta(\gamma) = \int_0^t \left[e^{-2\varphi(\gamma(\tau))}g(\dot\gamma(\tau), \dot\gamma(\tau))\right]^{\frac{1}{2}}d\tau = \int_0^t e^{-\varphi(\gamma(\tau))}d\tau.$$

But $\varphi(\gamma(\tau)) \geq d(\gamma(\tau), x_0) = \tau$, $e^{-\varphi(\gamma(\tau))} \leq e^{-\tau}$, $\int_0^t e^{-\varphi(\gamma(\tau))}d\tau \leq \int_0^t e^{-\tau}d\tau \leq 1$.

Therefore, $L_\delta(\gamma) \leq 1$, which shows that $\delta(x_0, x) \leq 1$, and X has diameter less than 2 for the distance δ.

5.7. Exercise 1.7. If X is compact, the first part of the following proof is of no use.

Consider Ω', Ω'' open, relatively compact, $cl(\Omega) \subset \Omega'$, $cl(\Omega') \subset \Omega'' \subset X$. Fix $\bar{K} \subset Z$, $\bar{K}$ compact. Set $V = U \times (I_B)^{(r-1)d_u}$. Then, $\Phi_r^\Sigma(\Omega' \times V)$ is relatively compact. Let $y_0 \notin \Phi_r^\Sigma(cl(\Omega') \times V)$.

First, we consider $\tilde{\Phi}$, smooth, equal to Φ_r^Σ on $\Omega' \times V$, $\tilde{\Phi}(x, \underline{u}_r) = y_0$ for $(x, \underline{u}_r) \notin \Omega'' \times V$. This is possible. Second, we will modify slightly this mapping $\tilde{\Phi}$, to get a mapping $\bar{\Phi}$, according to the following lemma.

Lemma 5.1. *($d_y > d_u$, $r > 2n$). There exists a smooth mapping $\bar{\Phi} : X \times V \to R^{rd_y}$, arbitrarily close (Whitney) to $\tilde{\Phi}$, such that*

(a) $\bar{\Phi} = \tilde{\Phi}$ *on* $(\Omega' \times V) \cup ((X \backslash \Omega'') \times V)$,
(b) $\bar{\Phi}((\Omega'' \backslash cl(\Omega')) \times V) \cap \bar{\Phi}(cl(\Omega') \times V) = \varnothing$.

Proof. The proof can be obtained using the transversality theorems. Let us give a direct proof here.

Let $\psi : X \to R_+$, smooth, $\psi = 0$ on $cl(\Omega') \cup (X \backslash \Omega'')$, $\psi > 0$ on the (relatively compact) complement $\mathcal{O}$ of this set, $\mathcal{O} = \Omega'' \backslash cl(\Omega')$. Consider the equation in $(x, \Lambda, \tilde{x}, \underline{u}_r)$ on $\mathcal{O} \times R^{rd_y} \times cl(\Omega') \times V$:

$$\psi(x)\Lambda + \tilde{\Phi}(x, \underline{u}_r) - \tilde{\Phi}(\tilde{x}, \underline{u}_r) = 0. \tag{196}$$

This equation has only the following trivial solution in Λ:

$$\Lambda = \frac{1}{\psi(x)}(\tilde{\Phi}(\tilde{x}, \underline{u}_r) - \tilde{\Phi}(x, \underline{u}_r)),$$

$$\Lambda : \mathcal{O} \times cl(\Omega') \times V \to R^{rd_y},$$

$$\Lambda : (x, \tilde{x}, \underline{u}_r) \to \Lambda(x, \tilde{x}, \underline{u}_r).$$

The mapping Λ is smooth. Because $d_y > d_u$ and $r > 2n$, $\dim(\mathcal{O} \times \Omega' \times V) < r\, d_y$, and $R^{rd_y} \backslash \mathrm{Im}(\Lambda)$ is dense (Sard's theorem). Take any $\bar{\Lambda}$ in this dense set, and set:

$$\bar{\Phi}(x, \underline{u}_r) = \psi(x)\bar{\Lambda} + \tilde{\Phi}(x, \underline{u}_r).$$

In fact, $\bar{\Phi}$ can be chosen in any Whitney neighborhood of $\tilde{\Phi}$, because $\bar{\Lambda}$ can be taken arbitrarily small, and ψ is compactly supported.

For $\bar{\Lambda}$, Equation (196) has no solution $(x, \tilde{x}, \underline{u}_r) \in \mathcal{O} \times cl(\Omega') \times V$. This means that $\bar{\Phi}(x, \underline{u}_r)$ is never equal to $\bar{\Phi}(\tilde{x}, \underline{u}_r)$ for $(x, \tilde{x}, \underline{u}_r) \in \mathcal{O} \times cl(\Omega') \times V$. ■

For $(x, \tilde{x}, \underline{u}_r) \in (X\backslash\Omega'') \times cl(\Omega') \times V$, $\bar{\Phi}(x, \underline{u}_r) \neq \tilde{\Phi}(\tilde{x}, \underline{u}_r)$ because $\bar{\Phi}(x, \underline{u}_r) = y_0$, and $y_0 \notin \Phi_r^\Sigma(cl(\Omega') \times V)$.

Finally,

$$\bar{\Phi}(x, \underline{u}_r) \neq \tilde{\Phi}(\tilde{x}, \underline{u}_r) = \Phi_r^\Sigma(\tilde{x}, \underline{u}_r),$$
$$\text{for } (x, \tilde{x}, \underline{u}_r) \in (X\backslash cl(\Omega')) \times cl(\Omega') \times V. \tag{197}$$

Moreover, $\bar{\Phi}$ is globally Lipschitz w.r.t. x, uniformly w.r.t. $\underline{u}_r$, and equal to Φ_r^Σ on $cl(\Omega') \times V$.

We have to show that

$$d(\eta_0, x_0) \leq \gamma \|\bar{\Phi}(\eta_0, \underline{u}_r) - \bar{\Phi}(x_0, \underline{u}_r)\|,$$
$$\text{for all } \eta_0 \in \bar{K}' \subset X, \ x_0 \in \Omega, \ \underline{u}_r \in V. \tag{198}$$

Here $\bar{K}' \subset X$ is a fixed compact subset of X. Typically, $\bar{K}'$ will be as $cl(K''')$ in the proof of Lemma 1.1 of Chapter 6:

$$\eta_0 = \eta(0, z_0) = \mathcal{H}(z_0, u_0(0), \tilde{u}_r(0), h(x_0, u_0(0))),$$

takes it values in $\bar{K}'$, when (z_0, x_0) takes values in $\bar{K} \times \Omega$, where $\bar{K}$ is a given compact subset of Z.

If $\eta_0 \in \bar{K}'\backslash\Omega'$ (which is compact), then

$$M \geq \|\bar{\Phi}(\eta_0, \underline{u}_r) - \bar{\Phi}(x_0, \underline{u}_r)\| \geq \delta > 0,$$

because, by the previous lemma, the minimum cannot be zero: We consider $\eta_0 \in \bar{K}'\backslash\Omega'$, $x_0 \in cl(\Omega)$ (two compact sets). The minimum will be reached, and if it is zero, it cannot be reached for $\eta_0 \in (X\backslash cl(\Omega'))$, by (197), and it cannot be reached for $\eta_0 \in \partial\Omega'$, because $S\Phi_r^\Sigma$ is an injective immersion.

Hence, if $\eta_0 \in \bar{K}'\backslash\Omega'$,

$$d(\eta_0, x_0) \leq L \leq \frac{L}{\delta}\|\bar{\Phi}(\eta_0, \underline{u}_r) - \bar{\Phi}(x_0, \underline{u}_r)\|.$$

The only thing that remains to be proved is therefore

$$d(\eta_0, x_0) \leq \gamma \|\bar{\Phi}(\eta_0, \underline{u}_r) - \bar{\Phi}(x_0, \underline{u}_r)\|$$
$$\leq \gamma \|\Phi_r^\Sigma(\eta_0, \underline{u}_r) - \Phi_r^\Sigma(x_0, \underline{u}_r)\|,$$
$$\text{for } (\eta_0, x_0, \underline{u}_r) \in \Omega' \times \Omega \times V. \tag{199}$$

This can be rewritten in a simpler way, by considering the distance δ on $X \times R^{rd_u}$, associated to the Riemannian metric over $X \times R^{rd_u}$:

$$g'((\dot{x}, \dot{\underline{u}}_r), (\dot{x}, \dot{\underline{u}}_r)) = g(\dot{x}, \dot{x}) + \|\dot{\underline{u}}_r\|^2_{R^{rd_u}}.$$

(Here, g is the Riemannian metric on X, associated to d.)

Then, (199) is implied by

$$d_{g'}((\eta_0, \underline{u}_r), (x_0, \underline{u}_r)) \leq \gamma \left\| S\Phi_r^{\Sigma}(\eta_0, \underline{u}_r) - S\Phi_r^{\Sigma}(x_0, \underline{u}_r)\right\|_{R^{r(d_y+d_u)}},$$
$$(\eta_0, x_0, \underline{u}_r) \in \Omega' \times \Omega' \times V. \tag{200}$$

Obviously, this is a consequence of the following general fact:

Lemma 5.2. *Let $\Phi : X \to R^p$ be an injective immersion, d_g a Riemannian distance on X, $K \subset X$ a compact. In restriction to K:*

$$d_g(x, y) \leq l\|\Phi(x) - \Phi(y)\|_{R^p}.$$

Proof. As in Exercise 1.5 of the same chapter, the question is local:

Let U be any open neighborhood of the diagonal in $X \times X$, V the complement of U in $X \times X$, and $W = (K \times K) \cap V$. Set $m = \inf_{(x,y)\in W} \|\Phi(x) - \Phi(y)\|$. $m > 0$ because W is compact, and Φ is an injective immersion. For $(x, y) \in W$, $d_g(x, y) \leq M = \frac{M}{m} m \leq \frac{M}{m}\|\Phi(x) - \Phi(y)\|_{R^p}$.

To solve the local problem, let us fix $x_0 \in X$, $y_0 = \Phi(x_0) \in R^p$, and consider a local diffeomorphism $\psi : (R^p, y_0) \to (R^p, 0)$, mapping $\Phi(X)$ onto a coordinate plane $P = \{x_{k+1} = 0, \ldots, x_p = 0\}$ (locally).

Now, $\|\psi \circ \Phi(x) - \psi \circ \Phi(y)\|_{R^p}$ is again a distance δ on a neighborhood $U_{x_0} \subset X$ of x_0, which is Riemannian because P is totally geodesic in $(R^p, \|.\|_{R^p})$.

Then, locally, we can write

$$d_g(x, y) \leq k_1\delta(x, y) = k_1\|\psi \circ \Phi(x) - \psi \circ \Phi(y)\|_{R^p}$$
$$\leq k_1k_2\|\Phi(x) - \Phi(y)\|_{R^p}.$$

Covering the diagonal of $K \times K$ by such neighborhoods $U_{x_0} \times U_{x_0}$, and extracting a finite covering gives the result. ■

In particular, we have proven that

$$d_g(\eta_0, x_0) \leq \gamma\|\bar{\Phi}(\eta_0, \underline{u}_r) - \bar{\Phi}(x_0, \underline{u}_r)\|,$$
$$\text{for all } (\eta_0, x_0, \underline{u}_r) \in \bar{K}' \times \Omega \times V.$$

Going back to the statement of the exercise, we get

$$\|\bar{\Phi}(\eta(t, z_0), \underline{u}_r(t)) - \bar{\Phi}(x(t, x_0), \underline{u}_r(t))\| \leq \lambda\gamma k(\alpha)e^{-\alpha t}\|\bar{\Phi}(\eta(0, z_0), \underline{u}_r(0)) - \bar{\Phi}(x_0, \underline{u}_r(0))\|,$$

where $\lambda\gamma$ depends on the compact $\bar{K}$, and for all $z_0 \in \bar{K}$, $x_0 \in \Omega$, $u \in U^{r,B}$. This shows that the modified observer is an exponential $U^{r,B}$ output observer, for $N = r$.

5.8. Exercise 2.1. Let us consider $\varphi : \tilde{\Gamma} = \Gamma \times V_B \subset R^{n+d_u} \to R$, with $\varphi = \varphi(x_1, \ldots, x_n, u)$. Here, we take φ equal to h or to one of the f_i's in the normal form (20). Let $z = (x_{r+1}, \ldots, x_n)$, $x = x_r$, $y = (x_1, \ldots, x_{r-1}, u_1, \ldots, u_{d_u})$. $\varphi(x, y, z)$ is an analytic function on $\tilde{\Gamma}$, such that

$$\begin{aligned} &\varphi \text{ does not depend on } z, \\ &\frac{\partial \varphi}{\partial x} > 0. \end{aligned} \tag{201}$$

$\tilde{\Gamma}$ is compact, convex (so are Γ and V_B). By analyticity, we can extend φ to an open, relatively compact neighborhood $V \subset R^m$ of $\tilde{\Gamma}$ ($m = n + d_u$), such that

$$\begin{aligned} &(i)\ \frac{\partial \varphi}{\partial x} > 0 \text{ on } V, \\ &(ii)\ \frac{\partial \varphi}{\partial z} = 0 \text{ on } V. \end{aligned} \tag{202}$$

Lemma 5.3. *There exists two open, relatively compact neighborhoods V', V'' of $\tilde{\Gamma}$, $cl(V') \subset V''$, $cl(V'') \subset V$, V' and V'' convex.*

Proof. Let us find only V', and repeat the proof with $\tilde{\Gamma} = V'$, to find V''. Consider $\hat{\Gamma}_n = \tilde{\Gamma} \cup \{B(z, \frac{1}{n}) | z \in \partial\tilde{\Gamma}\}$, where $B(z, \frac{1}{n})$ is the closed ball centered on z, with radius $\frac{1}{n}$. Set $V_n = Convex\ Hull(\hat{\Gamma}_n)$. We **claim** that for n sufficiently large, $V_n \subset V$.

Then, we just take $V' = Int(V_n)$ for n sufficiently large.

If the claim is false, there is a sequence $z_n \in V_n$, $z_n \notin V$, $z_n = t_n x_n + (1 - t_n) y_n$, $x_n, y_n \in \hat{\Gamma}_n$, $t_n \in [0, 1]$.

By compactness, we can assume that $z_n \to z$, $t_n \to t$, $x_n \to x$, $y_n \to y$, $x, y \in \tilde{\Gamma}$. Therefore, $z \in \tilde{\Gamma}$. It follows that $z_n \in V$ for n large, which is a contradiction. ∎

Lemma 5.4. *On V'' (and on V'), φ does not depend on z.*

Proof. For $(x, y, z), (x, y, z') \in V''$, by the convexity of V'',

$$\begin{aligned} \varphi(x, y, z') &= \varphi(x, y, z) + \sum (z'_i - z_i) \int_0^1 \frac{\partial \varphi}{\partial z_i}(x, y, tz' + (1 - t)z) dt, \\ &= \varphi(x, y, z). \end{aligned}$$ ∎

Now, if $\Pi : V'' \to R^{1+p}$ is the projection $(x, y, z) \to (x, y)$, with $p = r + d_u - 1$, then, $\Pi V'$ and $\Pi V''$ are convex and open,

$$\Pi\tilde{\Gamma} \subset \Pi V', \ cl(\Pi V') \subset \Pi V'',$$

and φ defines an analytic function $\tilde{\varphi}$ on $\Pi V''$:

$$\tilde{\varphi}(x, y) = \varphi(x, y, z) \text{ for } z \in \Pi^{-1}(x, y),$$

$$\frac{\partial \tilde{\varphi}}{\partial x} > 0, \text{ for } (x, y) \in \Pi V''.$$

Let $\Pi' : R^{1+p} \to R^p$ be the projection $\Pi'(x, y) = y$.

Lemma 5.5. *There exists a smooth mapping $s : R^p \to R$, compactly supported,*

$$(s(y), y) \in \Pi V', \textit{ for } y \in \Pi'\Pi\tilde{\Gamma}.$$

Proof. For all $y_0 \in \Pi'\Pi V'$, we chose $(x_0, y_0) \in \Pi V'$. Consider neighborhoods $U_{y_0} \subset \Pi'\Pi V'$, $V_{x_0} \times U_{y_0} \subset \Pi V'$ and the mapping

$$s_{y_0} : U_{y_0} \to V_{x_0} \times U_{y_0},$$
$$s_{y_0}(y) = x_0.$$

We also set $U_0 = R^p \backslash \Pi'\Pi\tilde{\Gamma}$, $s_0 = 0$.

The sets $\{U_0\} \cup \{U_{y_0} | y_0 \in \Pi'\Pi V'\}$ form an open covering of R^p. Let $\{(U_i', \chi_i) \,|i \in I\}$ be a partition of unity on R^p, $U_i' \subset U_0$, or $U_i' \subset U_{y_i}$ for some $y_i \in \Pi'\Pi V'$. If $U_i' \subset U_0$, we set $s_i = s_0 = 0$. If $U_i' \subset U_{y_i}$ (we select one), we set $s_i = s_{y_i}$. We consider

$$s : R^p \to R,$$
$$s(y) = \sum_{i \in I} \chi_i(y) s_i(y).$$

If $y \in \Pi'\Pi\tilde{\Gamma}$, by convexity of $\Pi V'$, $(s(y), y) \in \Pi V'$. ∎

With standard arguments, the function $\tilde{\varphi}$, defined on the open set $\Pi V''$, can be extended smoothly to all of R^{1+p}, in such a way that it is compactly supported and unchanged on $cl(\Pi V')$. Let us call the resulting function $\tilde{\varphi}$ again. Set

$$\bar{\varphi}(x, y) = \tilde{\varphi}(s(y), y) + \int_{s(y)}^{x} F \circ \frac{\partial \tilde{\varphi}}{\partial x}(\theta, y) d\theta,$$

where $F : R \to R$ is as follows:

F is increasing and locally constant outside a compact, equal to the identity on the interval $[m, M]$, equal to $\frac{m}{2}$ at $-\infty$, equal to 2 M at $+\infty$, with

$$m = \inf_{(x,y) \in cl(\Pi V')} \frac{\partial \tilde{\varphi}}{\partial x}(x, y), \; m > 0,$$

$$M = \max_{(x,y) \in cl(\Pi V')} \frac{\partial \tilde{\varphi}}{\partial x}(x, y).$$

Let us show that the function $\bar{\varphi}$ gives a solution of the exercise.

1. $\frac{\partial \bar{\varphi}}{\partial x}(x, y) = F \circ \frac{\partial \tilde{\varphi}}{\partial x}(x, y)$, hence, $\frac{m}{2} \leq \frac{\partial \bar{\varphi}}{\partial x} \leq 2M$.
2. $\frac{\partial \bar{\varphi}}{\partial y_i}(x, y) = \frac{\partial}{\partial y_i}(\tilde{\varphi}(s(y), y) + \frac{\partial}{\partial y_i}(\int_{s(y)}^{x} F \circ \frac{\partial \tilde{\varphi}}{\partial x}(\theta, y)d\theta) = (I) + (II)$.
   (I) is bounded, because $\tilde{\varphi}$ is compactly supported.
   $(II) = -\frac{\partial s}{\partial y_i}(y)F \circ \frac{\partial \tilde{\varphi}}{\partial x}(s(y), y) + \int_{s(y)}^{x} \frac{\partial}{\partial y_i}(F \circ \frac{\partial \tilde{\varphi}}{\partial x}(\theta, y))d\theta = (III) + (IV)$.
   (III) is bounded because $\frac{\partial s}{\partial y_i}(y)$ is bounded (s is compactly supported), and $\frac{m}{2} \leq F \leq 2M$.
   $(IV) = \int_{s(y)}^{x} F'(\frac{\partial \tilde{\varphi}}{\partial x}(\theta, y))\frac{\partial^2 \tilde{\varphi}}{\partial x \partial y_i}(\theta, y)d\theta$. The functions F' and $\frac{\partial^2 \tilde{\varphi}}{\partial x \partial y_i}$ are bounded by b, and $\frac{\partial^2 \tilde{\varphi}}{\partial x \partial y_i}$ is zero for $|\theta| \geq \theta_M$, because $\tilde{\varphi}$ is compactly supported.
   Hence, $(IV) \leq \int_{-\theta_M}^{\theta_M} b^2 d\theta = 2b^2\theta_M$.
3. $\bar{\varphi}$ coincides with $\tilde{\varphi}$ on $\Pi\tilde{\Gamma}$: for $(x, y) \in \Pi\tilde{\Gamma}$,

$$\bar{\varphi}(x, y) = \bar{\varphi}(s(y), y) + \int_{s(y)}^{x} \frac{\partial \bar{\varphi}}{\partial x}(\theta, y)d\theta,$$

but $\bar{\varphi}(s(y), y) = \tilde{\varphi}(s(y), y)$ by definition of $\bar{\varphi}$. Because $(s(y), y)$ and $(x, y) \in \Pi V'$, and $\Pi V'$ is convex, $(\theta, y) \in \Pi V'$ for all $\theta \in [s(y), x]$, and $\frac{\partial \bar{\varphi}}{\partial x}(\theta, y) = F \circ \frac{\partial \tilde{\varphi}}{\partial x}(\theta, y) = \frac{\partial \tilde{\varphi}}{\partial x}(\theta, y)$, by definition of F. Therefore, $\bar{\varphi}(x, y) = \tilde{\varphi}(x, y)$.

5.9. Exercise 2.2. This is a very elementary standard result from linear control theory.

5.10. Exercise 2.3. This is a standard result from stability theory (of linear differential equations).

5.11. Exercise 2.4. Let us consider the system on $Gl_+(m, R)$, $m = Np$:

$$\begin{aligned}\frac{d\Psi_u(t, s)}{dt} &= A_{N,p}\Psi_u(t, s) + \sum_{k,l=0}^{N-1} \sum_{i,j=1}^{p} \{u_{kp+i,lp+j}e_{kp+i,lp+j}\Psi_u(t, s)\}_{l \leq k},\\ &= A_u(t)\Psi_u(t, s),\\ \Psi_u(s, s) &= Id,\end{aligned}$$

as in Formula (92), where $|u_{i,j}(t)| \leq B$. Recall that $\Psi_u(t_1, t_2)\Psi_u(t_2, t_3) = \Psi_u(t_1, t_3)$. Set $\Psi_u(t, s) = [\Phi_u^{-1}]'(t, s)$. Then,

$$\begin{aligned}\frac{d\Phi_u(t, s)}{dt} &= -A_u(t)'\Phi_u(t, s),\\ \Phi_u(s, s) &= Id.\end{aligned}$$

Fix $T > 0$. Set $G_u = \int_0^T \Phi_u(T,s)C'C\Phi_u'(T,s)ds$. Here, G_u is the matrix under consideration in Lemmas 2.12–2.17 of Chapter 6.

$$G_u = \int_0^T (\Psi_u^{-1})'(T,s)C'C\Psi_u^{-1}(T,s)ds$$
$$= \int_0^T \Psi_u{}'(s,T)C'C\Psi_u(s,T)ds.$$

Hence, we have to consider the system:

$$\frac{d\Psi_u(s,T)}{ds} = A_u(s)\Psi_u(s,T),$$
$$\Psi_u(T,T) = Id.$$

This system is control affine. Hence, we can apply Lemma 2.10. (In fact it is a system with terminal condition on $[0,T]$, but the lemma also applies.)

Here, let us identify $L^\infty_{([0,T],R^l)}$ with $(L^\infty_{([0,T],R)})^l$, and endow $L^\infty_{([0,T],R)}$ with the weak-* topology. The subset $U_B = \{u = (u_1,\dots,u_l) | \sup_i |u_i|_\infty \leq B\} \subset L^\infty_{([0,T],R^l)}$ is compact.

The mapping

$$L^\infty_{([0,T],R^l)} \to C^0_{([0,T],Gl_+(m,R))},$$
$$u \to \Psi_u(s,T)$$

is continuous, by Lemma 2.10. Also, the mapping

$$C^0_{([0,T],Gl_+(m,R))} \to S_m,$$
$$\Psi(.) \to \int_0^T \Psi'(t)C'C\Psi(t)dt$$

is continuous (Lebesgue's dominated convergence). Hence, the mapping

$$\mathcal{F} : L^\infty_{([0,T],R^l)} \to S_m,\ u \to G_u$$

maps the compact set U_B onto a compact subset of S_m. In fact, $\mathcal{F}$ maps U_B onto a compact subset of the set of positive semi definite matrices, by definition. It is sufficient to prove that all elements in $\mathcal{F}(U_B)$ are positive definite (i.e., $\mathcal{F}$ maps U_B onto a compact subset of $S_m(+)$). This will imply Lemma 2.11. Assume that for $u^* \in L^\infty_{([0,T],R^l)}$, G_{u^*} is not positive definite. Let $x_0 \in Ker(G_{u^*})$, $x_0 \neq 0$. Then, $z(s) = \Psi_u(s,T)x_0$ is a solution of

$$\frac{dz(s)}{ds} = A_u(s)z(s),$$
$$z(T) = x_0,$$
$$Cz(s) = 0, \text{ for all } s \in [0,T].$$

This contradicts the observability.

5.12. *Exercise 2.5.*

1. Let $\tilde{b}_i$ denote the ith p-block component of $\tilde{b}$, and z_i denote the ith p-block component of z; $\tilde{b}_i(z) = \frac{1}{\theta^{i-1}} b_i(z_1, \theta z_2, \ldots, \theta^i z_i)$.

$$\begin{aligned}\|\tilde{b}_i(z) - \tilde{b}_i(w)\| &= \frac{1}{\theta^{i-1}} \big\| b_i\big(z_1, \theta z_2, \ldots, \theta^{i-1} z_i\big) \\ &\quad - b_i\big(w_1, \theta w_2, \ldots, \theta^{i-1} w_i\big)\big\| \\ &\leq \frac{L_b}{\theta^{i-1}} \big\| \big(z_1, \theta z_2, \ldots, \theta^{i-1} z_i\big) \\ &\quad - \big(w_1, \theta w_2, \ldots, \theta^{i-1} w_i\big)\big\| \\ &\leq L_b \left\| \left(\frac{1}{\theta^{i-1}} z_1, \frac{1}{\theta^{i-2}} z_2, \ldots, z_i\right) \right. \\ &\quad \left. - \left(\frac{1}{\theta^{i-1}} w_1, \frac{1}{\theta^{i-2}} w_2, \ldots, w_i\right) \right\| \\ &\leq L_b \|z - w\|,\end{aligned}$$

because $\theta > 1$.

2. With $i \times i$ block notations for b^* and $\tilde{b}^*$, $\tilde{b}^*_{i,j}(z) = \frac{1}{\theta^{i-1}} b^*_{i,j}(\Delta^{-1} z)\theta^{j-1}$, and b^* is lower block triangular. Hence, $\tilde{b}_{i,j} = 0$ for $j > i$; $\tilde{b}^*_{i,j}(z) = \theta^{(j-i)} b^*_{i,j}(\Delta^{-1} z)$, $j \leq i$, and $\theta > 1$. Hence, $\|\tilde{b}^*_{i,j}(z)\| \leq \|b^*_{i,j}(\Delta^{-1} z)\|$.

5.13. *Exercise 2.6.* $(Q + \lambda Q C' C Q)^{-1}$ is well defined for $|\lambda|$ small.

Let us show that

$$(I) = (Q + \lambda Q C' C Q)(Q^{-1} - C'(\lambda^{-1} + C Q C')^{-1} C) = Id.$$

Here, C is a linear form, and

$$(I) = (Q + \lambda Q C' C Q)\left(Q^{-1} - C' C \frac{\lambda}{1 + \lambda C Q C'}\right) = Id + (II).$$

$$(II) = \lambda Q C' C - (Q C' C \lambda + \lambda^2 Q C' C Q C' C) \frac{1}{1 + \lambda C Q C'},$$

$$\begin{aligned}(II) &= \frac{1}{1 + \lambda C Q C'} (\lambda Q C' C + \lambda^2 Q C' (C Q C') C - \lambda Q C' C \\ &\quad - \lambda^2 Q C' C Q C' C) \\ &= 0.\end{aligned}$$

Bibliography

[1] R. Abraham and J. Robbin, *Transversal Mappings and Flows*, W.A. Benjamin, Inc., 1967.

[2] V. I. Arnold, S. M. Gusein-Zade, and A. N. Varchenko, Singularités des applications différentiables, French translation, ed. MIR Moscou, Vol. I, II, 1986.

[3] M. Balde and P. Jouan, Observability of control affine systems, *ESAIM/COCV* 3, pp. 345–359, 1998.

[4] E. Bierstone and P. Milman, Extensions and liftings of C^∞ Whitney fields, *L'Enseignement Mathématique*, T.XXVIII, fasc. 1–2, 1977.

[5] E. Bierstone and P. Milman, Semi-analytic and subanalytic sets, *Publications de l'IHES*, No. 67, pp. 5–42, 1988.

[6] E. Bierstone and P. Millman, Geometric and differential properties of subanalytic sets, *Annals of Maths* 147, pp. 731–785, 1998.

[7] N. Bourbaki, Eléments de Mathématiques, Topologie générale, livre III, *Actualités Scientifiques et Industrielles*, 1142, Hermann, Paris, 1961.

[8] R. Bucy and P. Joseph, Filtering for stochastic processes with applications to guidance, Chelsea Publishing Company, 1968; second edition, 1987.

[9] J. Carr, Applications of centre manifold theory, *Appl. Math. Sci.* 35, Springer-Verlag, 1981.

[10] H. Cartan, Variétés analytiques réelles et variétés analytiques complexes, *Bulletin de la Société Mathématique de France* 85, pp. 77–99, 1957.

[11] H. Cartan, Variétés analytiques complexes et cohomologie, Colloque sur les fonctions de plusieurs variables, Bruxelles, pp. 41–55, 1953. Also in Collected Works of H. Cartan, Vol II, pp. 669–683, Springer-Verlag, 1979.

[12] F. Deza, Contribution to the synthesis of exponential observers, Ph.D. thesis, INSA de Rouen, France, June 1991.

[13] Z. Denkowska and K. Wachta, La sous analyticité de l'application tangente, *Bulletin de l'Académie Polonaise des Sciences*, XXX, No. 7–8, 1982.

[14] F. Deza, E. Busvelle, and J-P. Gauthier, High-gain estimation for nonlinear systems, *Systems and Control Letters* 18, pp. 295–299, 1992.

[15] M. Fliess and I. Kupka, A finiteness criterion for nonlinear input-output differential systems, *SIAM Journal Contr. and Opt.* 21, pp. 721–728, 1983.

[16] J-P. Gauthier, H. Hammouri, and I. Kupka, Observers for nonlinear systems; IEEE CDC Conference, Brighton, England, pp. 1483–1489, December 1991.

[17] J-P. Gauthier, H. Hammouri, and S. Othman, A simple observer for nonlinear systems, *IEEE Trans. Aut. Control* 37, pp. 875–880, 1992.

[18] J-P. Gauthier and I. Kupka, Observability and observers for nonlinear systems, *SIAM Journal on Control*, Vol. 32, No. 4, pp. 975–994, 1994.

[19] J. P. Gauthier and I. Kupka, Observability for systems with more outputs than inputs, *Mathematische Zeitschrift* 223 pp. 47–78, 1996.

[20] M. Goresky and R. Mc Pherson, *Stratified Morse Theory*, Springer-Verlag, 1988.

[21] F. Guaraldo, P. Macri, and A. Tancredi, Topics on real analytic spaces, Advanced Lectures in Maths, Friedrich Vieweg and Sohn, Braunschweig, 1986.

[22] H. Grauert, On Levi's problem and the imbedding of real analytic manifolds, *Annals of Math.*, 68(2), pp. 460–472, Sept. 1958.

[23] R. Hardt, Stratification of real analytic mappings and images, *Invent. Math.* 28, pp. 193–208, 1975.

[24] R. Hermann and all., Nonlinear controllability and observability, *IEEE Trans. Aut. Control* AC-22, pp. 728–740, 1977.

[25] H. Hironaka, *Subanalytic Sets, Number Theory, Algebraic Geometry and Commutative Algebra*, in honor of Y. Akizuki, Kinokuniya, Tokyo, No. 33, pp. 453–493, 1973.

[26] M. W. Hirsch, *Differential Topology*, Springer-Verlag, Graduate Texts in Math., 1976.

[27] L. Hörmander, *An Introduction to Complex Analysis in Several Variables*, North Holland Math. Library, Vol. 7, 1973.

[28] M. Hurley, Attractors, Persistance and density of their basis, *Trans. Am. Math. Soc.* 269, pp. 247–271, 1982.

[29] A. Jaswinsky, Stochastic Processes and Filtering Theory, Academic Press, New York, 1970.

[30] P. Jouan, Singularités des systèmes non linéaires, observabilité et observateurs, Ph.D. thesis, University of Rouen, 1995.

[31] P. Jouan, Observability of real analytic vector fields on a compact manifold, *Systems and Control Letters* 26, pp. 87–93, 1995.

[32] P. Jouan and J. P. Gauthier, Finite singularities of nonlinear systems. Output stabilization, observability and observers, *Journal of Dynamical and Control Systems* 2(2), pp. 255–288, 1996.

[33] L. Kaup and B. Kaup, Holomorphic functions of several variables, *De Gruyter Studies in Math.*, 1983.

[34] W. Klingenberg, Riemannian geometry, *De Gruyter Studies in Math.*, 1982.

[35] J. Kurzweil, On the inversion of Lyapunov's second theorem, On stability of motion, *Transl. Am. Math. Soc.* pp. 19–77, 1956.

[36] J. Lasalle and S. Lefschetz, *Stability by Lyapunov's Direct Method with Applications*, Academic Press, New York, 1961.

[37] S. Lefschetz, *Ordinary Differential Equations: Geometric Theory*, J. Wiley Intersciences, 1963.

[38] S. Lojasiewicz, Triangulation of semi analytic sets, *Annal. Sc. Nor. Sup. PISA*, pp. 449–474, 1964.

[39] D. G. Luenberger, Observers for multivariable systems, *IEEE Trans. Aut. Control* 11, pp. 190–197, 1966.

[40] J. L. Massera, Contribution to stability theory, *Annals of Math.* 64, pp. 182–206, 1956.

[41] J. W. Milnor, *Differential Topology*, Lectures on Modern Mathematics, T. L. Saaty, ed., Vol. II, Wiley, New York, 1964.

[42] J. W. Milnor, On the Concept of attractors: Corrections and remarks, *Comm. Math. Phys.* 102, pp. 517–519, 1985.

[43] R. Narasimhan, *Introduction to the Theory of Analytic Spaces*, Springer, Middelburg New York. Lecture Notes in Mathematics 25, 1966.

[44] N. Rouche, P. Habets and M. Laloy, *Stability Theory by Lyapunov's Direct Method*, Lecture Notes in Applied Mathematical Sciences 22, Springer-Verlag, New York, 1977.

[45] M. Shiota, *Geometry of Subanalytic and Semi-Algebraic Sets*, Birkhauser, P.M. 150, 1997.

[46] H. J. Sussmann, Trajectory regularity and real analyticity, some recent results, *Proceedings of 25th CDC Conference*, Athens, Greece, Dec. 1986.

[47] H. J. Sussmann, Single input observability of continuous time systems, *Mathematical Systems Theory*, 12(4), pp. 371–393, 1979.

[48] M. Tamm, Subanalytic sets in the calculus of variations, *Acta Mathematica* 146, 3.4, pp. 167–199, 1981.

[49] J. C. Tougeron, *Idéaux de fonctions différentiables*, Springer-Verlag, New York, 1972.

[50] J. L. Verdier, Stratifications de Whitney et théorème de Bertini-Sard, *Invent. Math.* 36, pp. 295–312, 1976.

[51] F. W. Wilson, Jr., The structure of the level surfaces of a Lyapunov function, *Journal Diff. Equ.* 3, pp. 323–329, 1967.

[52] H. Whitney, Analytic extensions of differentiable functions defined in closed sets, *Trans. Am. Math. Soc.* 36, pp. 63–89, 1934.

[53] O. Zariski and P. Samuel, *Commutative Algebra*, Van Nostrand Company, 1958.

[54] C. D. Holland, *Multicomponent Distillation*, Prentice Hall, Englewood Cliffs, NJ, 1963.

[55] H. H. Rosenbrock, A Lyapunov function with applications to some nonlinear physical systems, *Automatica* 1, pp. 31–53, 1962.

[56] J. Alvarez, R. Suarez, and A. Sanchez, Nonlinear decoupling control of free radical polymerization continuous stirred tank reactors, *Chem. Ing. Sci.* 45, pp. 3341–3357, 1990.

[57] D. K. Adebekun and F. J. Schork, Continuous solution polymerization reactor control, 1, *Ind. Eng. Chem, Res.* 28, pp. 1308–1324, 1989.

[58] D. Bossane, Nonlinear observers and controllers for distillation columns, Ph.D. thesis, University of Rouen, France, 1993.

[59] M. Van Dootingh, Polymerization radicalaire, commande géométrique et observation d'état à l'aide d'outils non-linéaires, Ph.D. thesis, University of Rouen, France, 1992.

[60] N. Petit, et al., Control of an industrial polymerization reactor using flatness, Second NCN Workshop, Paris, June 5–9, 2000, to appear in Lecture Notes in Control and Informatioon Sciences, Springer-Verlag.

[61] P. Rouchon, Dynamic simulation and nonlinear control of distillation columns (in French), Ph.D. thesis, Ecole des Mines de Paris, France, 1990.

[62] T. Takamatsu, I. Hashimoto, and Y. Nakai, A geometric approach to multivariable control system design of a distillation column, *Automatica* 15, pp. 178–202, 1979.

[63] F. Viel, Stabilité des systèmes non linéaires controlés par retour d'état estimé. Application aux réacteurs de polymérisation et aux colonnes à distiller, Ph.D. thesis, University of Rouen, France, 1994.

[64] F. Viel, E. Busvelle, and J-P. Gauthier, A stable control structure for binary distillation columns, *International Journal on Control* 67(4), pp. 475–505, 1997.

[65] F. Viel, E. Busvelle, and J-P. Gauthier, Stability of polymerization reactors using input-output linearization and a high gain observer, *Automatica* 31, pp. 971–984, 1995.

Index of Main Notations

Following is a list of general notations used throughout the book. Local symbols are not listed. In particular, any symbol relative to the applications in Chapter 8 is not included in the list.

$R, \mathcal{N}$ sets of real numbers and integers respectively
$[a, b[$ real semi open interval $\{x \in R | a \leq x < b\}$
$L_f h$ Lie derivative of the function h in the direction of the vector field f
$L_f^k h$ k times iterated Lie derivative
$\partial_j = \frac{\partial}{\partial u_j}$
$(\Sigma)\ \begin{array}{l} \frac{dx}{dt} = f(x, u), \\ y = h(x, u), \end{array}$ system, Chapter 1, Section 1
X state space, n-dimensional, idem
d_u, d_y dimensions of control space and output space, idem
$U = I^{d_u}$, $I \subset R$ closed interval, set of values of control, idem
$F:$ set of parametrized vector fields $\dot{x} = f(x, u)$, F^r, $r = 1, .., \infty, \omega$, idem
$H:$ set of output functions $h(x, u)$, H^r, $r = 1, .., \infty, \omega$, idem
$S = F \times H$: set of systems Σ, S^r, $r = 1, .., \infty, \omega$, idem
$L^\infty[U]$, $L[R^{d_y}]$ sets of control functions and output functions, Chapter 2, Section 1
$e(u, x)$ escape time, idem
P input-output mapping, idem
P_u *initial state* $\rightarrow$ *output* $-$ *trajectory* mapping, idem
$T\Sigma$ first variation of a system Σ, idem
dP tangent mapping to P, idem
$dP(\hat{u}, x)$ idem
$D(u) = \{D_0(u) \supset D_1(u) \supset \ldots \supset D_{n-1}(u)\}$ canonical flag, Chapter 2, Section 2
$\Phi_k^\Sigma : X \times U \times R^{(k-1)d_u} \rightarrow R^{kd_y}$, $\Phi_k^\Sigma : (x_0, u, u', \ldots, u^{(k-1)}) \rightarrow (y, y', \ldots, y^{(k-1)})$, Chapter 2, Section 3
$S\Phi_k^\Sigma : X \times U \times R^{(k-1)d_u} \rightarrow R^{kd_y} \times R^{kd_u}$,
$S\Phi_k^\Sigma : (x_0, u, u', \ldots, u^{(k-1)}) \rightarrow (y, y', \ldots, y^{(k-1)}, u, u', \ldots, u^{(k-1)})$ idem
$PH(k)$ phase variable property of order k, idem
$f^N(x, u^{(0)}, \ldots, u^{(N-1)})$, b^N Chapter 2, Section 4
Σ^N N^{th} dynamical extension of Σ, idem
$\underline{u}_N = (u^{(0)}, \ldots, u^{(N-1)})$ idem
$\underline{y}_N = (y^{(0)}, \ldots, y^{(N-1)})$ idem
Φ_k, $S\Phi_k$ idem

Θ^Σ observation space, Chapter 2, Section 5
Δ_Σ trivial distribution of Σ, idem
Ξ^Σ space of functions, relative to Σ, of the form
$L^{k_1}_{f_u}(\partial_{j_1})^{s_1} L^{k_2}_{f_u}(\partial_{j_2})^{s_2} \ldots\ldots L^{k_r}_{f_u}(\partial_{j_r})^{s_r} h_{i,u}$ idem
$\bar{\Delta}_\Sigma(u)$ idem
$Lie(\Sigma)$ Lie algebra of a system Σ, Chapter 2, Appendix
$A_\Sigma(x_0)$ Accessibility set of Σ through x_0, Chapter 2, Appendix
$O_\Sigma(x_0)$ Orbit of Σ through x_0, Chapter 2, Appendix
$d_X h, d_U h$ differentials of the mapping $h(x,u) : X \times U \to R^s$ with respect to the variable $x \in X$ only (resp. $u \in U$ only)
If $f_u = f(x,u) \in F$ is a parametrized vector field, $d_U f$ is a mapping: $TU \approx U \times R^{d_u} \to \kappa(X)$, the set of smooth vector fields over X, etc ... Chapter 3
$(\Sigma_A)\begin{cases} \dot{x} = f(x) + ug(x), \\ y = h(x), \end{cases}$ control affine system, Chapter 3, Section 4.2
$(\Sigma \times \Sigma)\begin{cases} \dot{x}_1 = f(x_1,u), \\ \dot{x}_2 = f(x_2,u), \end{cases}$ $y = h(x_1,u) - h(x_2,u)$, product of a system Σ by itself, idem
$(B)\begin{cases} \dot{x} = Ax + u(Bx+b), \\ y = Cx, \end{cases}$ $x \in R^n$, bilinear system, Chapter 3, Section 4.3
$S^{0,r}$ set of C^r systems, where h is independent of u, Chapter 4, Section 1
$H^{0,r}$ set of C^r output functions h that are independent of u, idem
B_H, B_{H^0}, B_F idem
J^kF, J^kH, J^kH^0 bundles of k-jets of C^r sections of B_F, B_H, B_{H^0}, idem
$\times_X$ fiber product of bundles over X, idem
$\underline{\times_X}$ idem
$\overline{J^kS}$ k-jets of systems, idem
ev_k evaluation mapping, idem
$S^K, S^{0,K}$ set of holomorphic systems over K, idem
$j^k\Sigma, j^kf, j^kh$ k-jets of Σ, f, h, idem
$T_X f(x)$ linearized of a vector field f at $x \in X$, $f(x) = 0$, idem
$D\Phi_N : ((X \times X)\setminus\Delta X) \times U \times R^{(N-1)d_u} \times S \to R^{Nd_y} \times R^{Nd_y}$ idem
$DS\Phi_N : ((X \times X)\setminus\Delta X) \times U \times R^{(N-1)d_u} \times S \to R^{Nd_y} \times R^{Nd_y} \times R^{Nd_u}$ idem
$D\Phi^\Sigma_N(x_1, x_2, \underline{u}_N) = D\Phi_N(x_1, x_2, \underline{u}_N, \Sigma)$,
$DS\Phi^\Sigma_N(x_1, x_2, \underline{u}_N) = DS\Phi_N(x_1, x_2, \underline{u}_N, \Sigma)$,
$\hat{\varphi} = j^\infty_t(\varphi) = \sum_{n=0}^{\infty} t^n \varphi_n$ infinite jet of φ w.r.t. t, Chapter 4, Section 7.2
$\underline{f}_N = (f^{(0)}, f^{(1)}, \ldots, f^{(N)})$ N-jet of a curve, Chapter 5, Section 1
$\tilde{f}_N = (f^{(1)}, \ldots, f^{(N)})$, $\underline{f}_N = (f^{(0)}, \tilde{f}_N)$
$\mathfrak{R}_N(x_0), \mathfrak{R}_N(x_0, \underline{u_0}_N), \hat{\mathfrak{R}}_N(x_0, \underline{u_0}_N)$ rings of germs, idem
$\mathfrak{R}_N, \mathfrak{R}_N, \hat{\mathfrak{R}}_N$ idem
$\Phi^\Sigma_{N,\bar{u}_{N-1}} : X \times U \to R^{Nd_y}$ restricted mapping, idem
$S\Phi^\Sigma_{N,\bar{u}_{N-1}} : X \times U \to R^{Nd_y} \times R^{d_u}$ suspended restricted mapping, idem
$\check{\mathfrak{R}}$ ring of analytic germs of functions of the form $G(u, \varphi_1, \ldots, \varphi_p)$, with $\varphi_i = L^{k_1}_{f_u}(\partial_{j_1})^{s_1} L^{k_2}_{f_u}(\partial_{j_2})^{s_2} \ldots\ldots L^{k_r}_{f_u}(\partial_{j_r})^{s_r} h$, $k_i, s_i \geq 0$, idem
$\check{\mathfrak{R}}_N$, same ring as $\check{\mathfrak{R}}$, with $\sum k_i + \sum s_i \leq N-1$ idem
$\mathfrak{R}\{u; u_0\}$ Chapter 5, Section 2
$ACP(N)$ ascending chain property, Chapter 5, Section 2

$\Sigma_{Oy}^{r,B,\Omega}$, $\Sigma_{Oye}^{r,B,\Omega}$ output observer (resp. exponential), Chapter 6, Section 1
$\Sigma_{Ox}^{r,B,\Omega}$, $\Sigma_{Oxe}^{r,B,\Omega}$ state observer (resp. exponential), idem
$\underline{x}_i = (x^0, \ldots, x^i)$ Chapter 6, Section 2
$A_{n,p}(np, n)$ block-antishift matrix:

$$A_{n,p} = \begin{pmatrix} 0, Id_p, 0, \ldots\ldots\ldots, 0 \\ 0, 0, Id_p, 0, \ldots\ldots, 0 \\ 0, \\ 0, 0, \ldots\ldots\ldots, 0, Id_p \\ 0, 0, \ldots\ldots\ldots\ldots, 0, 0 \end{pmatrix},$$

Chapter 6, Section 2.4.1
$\tilde{\mathfrak{R}}_N$ C^∞ analog of the ring $\hat{\mathfrak{R}}_N$, Chapter 7, Section 2
$\bar{\mathfrak{R}}_N$ C^∞ analog of the ring $\mathfrak{R}_N$, idem
$\tilde{\mathfrak{R}}$ C^∞ analog of the ring $\check{\mathfrak{R}}$, idem

Index